1985

INTRODUCTORY APPLIED PROBABILITY

ELLIS HORWOOD SERIES IN
MATHEMATICS AND ITS APPLICATIONS

Series Editor: **Professor G. M. BELL,** Chelsea College, University of London
(and within the same series)
Statistics and Operational Research
Editor: **B. W. CONOLLY,** Chelsea College, University of London ·

INTRODUCTORY
APPLIED PROBABILITY

G. P. BEAUMONT, B.Sc., M.A., M.Sc.
Senior Lecturer in Statistics
Department of Statistics and Computer Science
Royal Holloway College University of London

ELLIS HORWOOD LIMITED
Publishers · Chichester

Halsted Press: a division of
JOHN WILEY & SONS
New York · Brisbane · Chichester · Ontario

First published in 1983 by
ELLIS HORWOOD LIMITED
Market Cross House, Cooper Street, Chichester, West Sussex, PO19 1EB, England

*The publisher's colophon is reproduced from James Gillison's drawing of the
ancient Market Cross, Chichester.*

Distributors:

Australia, New Zealand, South-east Asia:
Jacaranda-Wiley Ltd., Jacaranda Press,
JOHN WILEY & SONS INC.,
G.P.O. Box 859, Brisbane, Queensland 4001, Australia

Canada:
JOHN WILEY & SONS CANADA LIMITED
22 Worcester Road, Rexdale, Ontario, Canada.

Europe, Africa:
JOHN WILEY & SONS LIMITED
Baffins Lane, Chichester, West Sussex, England.

North and South America and the rest of the world:
Halsted Press: a division of
JOHN WILEY & SONS
605 Third Avenue, New York, N.Y. 10016, U.S.A.

©1983 G. P. Beaumont/Ellis Horwood Ltd.

British Library Cataloguing in Publication Data
Beaumont, G. P.
Introductory applied probability. –
(Ellis Horwood series in mathematics and its applications)
1. Probabilities
I. Title
519   QA273

Library of Congress Card No. 83–10700

ISBN 0-85312-392-6 (Ellis Horwood Ltd. – Library Edn.)
ISBN 0-85312-650-X (Ellis Horwood Ltd. – Student Edn.)
ISBN 0-470-27481-6 (Halsted Press – Library Edn.)
ISBN 0-470-27473-5 (Halsted Press – Student Edn.)

Printed in Great Britain by R. J. Acford, Chichester.

Contents

6

Preface

The aim of this book is to provide direct access to a selection of important applications of probability, namely Birth and Death Processes, Queueing Theory, Renewal Theory, Reliability and Inventory Control. It attempts to find a middle way between the elaborate framework demanded by a course in Stochastic Processes and the sometimes diffuse outlines provided in handbooks of Operational Research. It is hoped that any loss of elegance and generality is compensated by gains in immediacy and focus.

The treatment is mathematical but not advanced. It presupposes a first-year course in Calculus and the usual amount of Probability provided in a first course in Statistics. The book is otherwise self-sufficient — starting with a review of certain indispensable standard distributions and the Poisson process and concluding with an appendix on the Laplace transform.

The needs of the average student have largely dictated the presentation. The steps in the derivations and examples have been set out in some detail with attention to those points which are habitual sources of difficulty. In this regard a special effort has been made with respect to Inventory Theory for which topic there are few *intermediate* discussions. Because of the relative unfamiliarity of some of the techniques employed in this area it was necessary to supply a considerable amount of detail in the illustrative examples including the numerical aspects. Many typical problems have been included — including a selection from recent examinations set by various universities. To all of these, at some cost to the author's blood-pressure, brief solutions and comments have been provided. An asterisk before a section or a problem indicates a more difficult piece of material.

Random variables are denoted by capital letters, such as X, and a value assumed by such a random variable by the corresponding lower case letter x. In discussing Inventory Theory, this convention involves a superabundance of type changes and increases the difficulty of perusing the equations. In Chapter 6, the reader is advised when this scheme is no longer strictly adhered to. The corresponding vectors are denoted by X and x.

We employ the term probability density function (p.d.f.) whether a random variable is continuous or discrete. The base e is not shown in logarithms, which are always natural. In summations, $\sum_{i=1}^{n} X_i$ is often shortened to $\sum_{1}^{n} X_i$.

9

Acknowledgements

I am indebted to the following sources for permission to publish:

The Senate of the University of London and the Universities of Birmingham, Brunel, Exeter, Hull, Leeds, Nottingham, Manchester, Reading, Sheffield, Southampton and Wales for questions from past examination papers. These institutions are in no way responsible for any of the solutions provided.

Most of the various drafts of the book have been scrutinized by the singularly beady eye of Professor K. Bowen. I am very grateful for the time he devoted to discussing the material. I was greatly assisted in the final stages by the breezy encouragement of Professor B. Conolly. Faintness of heart prevented me from rising to his more challenging suggestions. I also wish to thank Professor P. R. Freeman for reading and commenting on the introductory chapter. I am grateful to Mrs B. Rutherford for typing the manuscript and coping with demands for corrections with unfailing good humour.

CHAPTER 1

Introduction

1.1 RANDOM STREAM OF EVENTS

Attempts are frequently made to 'explain' the functioning of real systems through the construction of mathematical models. The ability of such models to predict the future behaviour of such systems can often be improved by incorporating probabilistic elements. In particular, taking account of possible random fluctuations may lead to more profitable decision making. We offer an oversimplified example which displays some typical features. Airlines know that some booked passengers will fail to turn up for a flight. If the flight is popular this behaviour will lead to loss of revenue. Such losses can be reduced by deliberately over-booking. Of course this tactic may lead to irritated intending passengers being refused a seat and subsequently demanding compensation! The question is, how much over-booking should be made?

Now, the airline almost certainly has records of the proportion of 'no-shows' for the flight in question. Thus the simplest remedy is to over-book to the extent that after the *expected* number of defaults, the aircraft is just full. But this solution is too static since it takes insufficient account of possible random fluctuations. It is better to estimate the probability distribution for the number of booked passengers turning up. In some situations, the probability of any overflow at all may be the crucial feature — as in the chance of an insurance firm going bankrupt. In the present instance, the penalty per passenger refused and the profit per seat occupied seem more appropriate measures. This suggests booking a number for which the *expected loss* is a minimum.

We shall be studying several processes in which probabilistic methods may be fruitfully employed. No model can reproduce all the complexities of a real phenomenon, so attention is restricted to some features of interest. Any observable feature will be called an event. Many of the processes we shall be studying evolve in time and we may conceive of events as streaming towards an observer. Some examples are: the arrivals of customers at a service point, the failures of a key part in a machine, a store running out of stock. Owing to random fluctuations outside our control, we are not in general able to predict the time of the next event. Where we have some control, the process may show deterministic features. Thus in a hospital appointment system, the time of arrival of the next patient may be uncertain, but the time at which he is seen may be controlled.

At the opposite extreme is the idea of a *completely random stream of events*. To sharpen this intuitive idea, we first look forward in time to the next occurrence in the stream of events. If this event may arrive at any time, then the wait is both positive and continuous. 'Completely random' suggests that if we have been observing the stream for some time, the past history of the process tells us nothing about its future. That is to say, knowledge of the number and occurrence times of previous events does not alter the distribution of times to ensuing events. In particular, suppose that the first such event has not occurred by time t_0, then the conditional probability of waiting at least a further time t is $[1-F(t+t_0)]/[1-F(t_0)]$; where F(.) is the cumulative distribution function of the time to the first event. If this equals the probability of waiting at least a further time t to the first event, we require

$$\frac{1-F(t + t_0)}{1-F(t_0)} = 1-F(t). \tag{1.1}$$

Differentiating both sides of equation (1.1) with respect to t,

$$\frac{-F'(t + t_0)}{1-F(t_0)} = -F'(t).$$

If we allow $t \to 0$ and require $F'(.)$ to be continuous, we must have

$$\frac{-F'(t_0)}{1-F(t_0)} = -F'(0),$$

$$\frac{d}{dt_0} [\log \{1-F(t_0)\}] = -F'(0),$$

$$\log \{1-F(t_0)\} = -F'(0)t_0 + k,$$

$$1-F(t_0) = \exp [1-F'(0)t_0 + k].$$

But, for a non-negative continuous random variable, $F(0) = 0$, hence $k = 0$. Finally

$$F(t_0) = 1-\exp [-F'(0) t_0]. \tag{1.2}$$

The corresponding probability density function, $f(.)$, of the time to the first event satisfies $f(t_0) = F'(t_0) = F'(0) \exp [-F'(0)t_0]$. This is of the form $f(t_0) = \lambda \exp(-\lambda t_0)$ which is the probability density function of an exponential distribution with parameter $\lambda = F'(0)$.

This kind of consideration motivates our definition of a *completely random stream of events, namely, that the times between each of successive members of the sequence have independent exponential distributions with a common mean.*

This common mean remains undetermined and corresponds to the idea that completely random streams of events may be more or less dense.

Looking forward in this way forces us to concede that the time to the next event is possibly infinite. We can re-state the problem so that only finite quantities are involved. This can be done by looking backwards from the current time t. Suppose we know that just one of the events has occured before t but not when. Then, if its arrival were completely random, we surely imply that its conditional distribution over $(0,t)$ is uniform. This corresponds to the primitive idea of picking a point in an interval 'at random'. To be sure, the probability of selecting a particular point is zero, but the probability of selecting a point from a sub-interval is proportional to the length of the sub-interval. It can be shown that if the time of arrival, T, satisfies

$$Pr[T < t_0 \mid \text{one event in } (0,t)] = t_0/t,$$

for any $t_0 < t$ and all t, then T necessarily has an exponential distribution. At this point it will be convenient to revise the properties of the exponential and related distributions.

1.2 EXPONENTIAL DISTRIBUTION

The continuous, non-negative random variable X, is said to have the exponential distribution, with positive parameter λ, if it has p.d.f.

$$f(x) = \lambda \exp(-\lambda x), x \geqslant 0.$$

Then

$$Pr(X \geqslant x_0) = \int_{x_0}^{\infty} \lambda \exp(-\lambda x)dx = \exp(-\lambda x_0). \tag{1.3}$$

Hence the cumulative distribution function, $F(.)$, satisfies $F(x_0) = Pr(X \leqslant x_0) = 1 - \exp(-\lambda x_0)$. (Compare with equation (1.2).) From (1.3) we have, for $x_0 < x_1$,

$$Pr(X \geqslant x_1 \mid X \geqslant x_0) = \frac{Pr(X \geqslant x_1)}{Pr(X \geqslant x_0)}$$

$$= \exp[-\lambda(x_1 - x_0)];$$

that is, if X is known to assume a value greater than x_0, then the probability that it assumes a value greater than $x_1 (> x_0)$ only depends on the difference $x_1 - x_0$ and not on the individual values x_0, x_1 at all. This is the basis for describing the exponential as a distribution without memory. The use of such a distribution in a practical setting thus entails consequences of the following type. If the times between the arrivals of taxis at a rank are exponential, then, however long the front customer has been there, his expectations as to the waiting time remaining are unaltered.

Many properties of the exponential distribution are easily obtained from its moment generating function (m.g.f.), $M_X(\theta)$. Now

$$M_X(\theta) = E[e^{X\theta}] = \int_0^\infty \exp(x\theta)\,\lambda \exp(-\lambda x)dx = \lambda/(\lambda-\theta), \quad (1.4)$$

where $|\theta|$ is restricted to be less than λ both to make the integral finite and to make the expansion of $\lambda/(\lambda-\theta)$ as a power series valid.

$$M_X(\theta) = (1-\theta/\lambda)^{-1} = \sum_{r=0}^\infty (\theta/\lambda)^r \text{ and } E(X) = 1/\lambda, E(X^2) = 2/\lambda^2.$$

Hence $V(X) = 1/\lambda^2$.

Problem 1. If X has the exponential distribution with parameter λ, confirm that $E(X) = 1/\lambda$, both by calculating $E(X)$ as $\int_0^\infty \lambda x \exp(-\lambda x)dx$ and also by evaluating $\dfrac{dM_X(\theta)}{d\theta}$ at $\theta = 0$. ■

Problem 2. If $X_1, X_2, \ldots, X_k$ is a random sample from the exponential distribution with parameter λ, show that Min $(X_1, X_2, \ldots, X_k)$ has another exponential distribution with parameter $k\lambda$.

Suppose $X_1, X_2, \ldots, X_k$ are the lives of k new lamps switched on simultaneously. Let T_i be the time to the ith lamp to fail and $Z_1 = T_1, Z_i = T_i - T_{i-1}$ $(i = 2, 3, \ldots, k)$. Explain why Z_i has an exponential distribution with parameter $[k-(i-1)]\lambda$ and deduce that

$$Pr\left(\sum_1^k Z_i < t\right) = [1-\exp(-\lambda t)]^k. \quad ■$$

The Gamma Distribution

We are also able to find the distribution of the time, Y, to the nth in a completely random stream of events, for this is but the sum of n independent random variables, each having the same exponential distribution. The m.g.f. of Y is

$$M_Y(\theta) = E[\exp(Y\theta)]$$

$$= E\left[\exp\left(\sum_1^n X_i\,\theta\right)\right]$$

$$= \prod_1^n E[\exp(X_i\,\theta)]$$

$$= \left(\frac{\lambda}{\lambda-\theta}\right)^n, \text{ from (1.4).} \quad (1.5)$$

It is readily verified that the distribution for which $[\lambda/(\lambda-\theta)]^n$ is the m.g.f. has p.d.f.

$$f(y) \;=\; \frac{\lambda\,(\lambda y)^{n-1}\exp{(-\lambda y)}}{(n-1)!}, y \geqslant 0. \tag{1.6}$$

The reader may recognise, from (1.6), that Y has a gamma distribution with parameters (n, λ). The basic properties of the gamma distribution are reviewed in the following problems.

Problem 3. We define the *gamma function*, $\Gamma(\alpha)$, for $\alpha > 0$, as

$$\Gamma(\alpha) = \int_0^\infty x^{\alpha-1}\,\exp{(-x)}dx.$$

(i)   Show that if $\alpha > 1$, $\Gamma(\alpha) = (\alpha-1)\Gamma(\alpha-1)$ and hence that, for a positive integer m, $\Gamma(m) = (m-1)!$

(ii) Show that $\Gamma(\alpha) = 2\int_0^\infty y^{2\alpha-1}\,\exp{(-y^2)}dy$. Deduce that $\Gamma(\frac{1}{2}) = \sqrt{\pi}$. ■

Problem 4. The continuous random variable X is said to have the *gamma distribution* with positive parameters (α, λ) if it has p.d.f.

$$f(x) \;=\; \frac{\lambda\,(\lambda x)^{\alpha-1}\exp{(-\lambda x)}}{\Gamma(\alpha)}, \; x \geqslant 0.$$

Show that for a gamma (α, λ) distribution:
(i) $E(X) = \alpha/\lambda$, $V(X) = \alpha/\lambda^2$, $E[\exp{(X\,\theta)}] = [\lambda/(\lambda-\theta)]^\alpha$; and
(ii) if $\alpha > 1$, $E(1/X) = \lambda/(\alpha-1)$, and that $f(x)$ has a mode at $x = (\alpha-1)/\lambda$.

Also show that if X_i has the gamma (α_i, λ) distribution $(i = 1,2,\ldots,n)$ and the X_i are independent then

(iii) the distribution of $\sum_1^n X_i$ has a gamma distribution with parameters

$$\left(\sum_1^n \alpha_i, \lambda\right) \;\; ■$$

The Poisson Distribution

Now that we have the distribution of the time to the nth event in a random stream, we can readily calculate the probability that just n of these events happen in the interval $(0,t)$. For this to be the case, the nth event must occur at some time, y, no later than t and the wait for the next event must be at least $t-y$. The total probability, is, using (1.6) and (1.3),

$$\int_0^t \frac{\lambda (\lambda y)^{n-1} \exp (-\lambda y) \exp [-\lambda (t-y)]}{(n-1)!} \, dy$$

$$= \exp (-\lambda t) \int_0^t \frac{\lambda (\lambda y)^{n-1}}{(n-1)!} \, dy$$

$$= \frac{(\lambda t)^n \exp (-\lambda t)}{n!} \qquad\qquad (1.7)$$

(Note: the waiting time is independent of any events before y.) The expression (1.7) is clearly the p.d.f. of a Poisson distribution with parameter λt. Strictly speaking we have covered the cases $n = 1,2, \ldots$. But the probability of no event in $(0,t)$ is $\exp(-\lambda t)$, and so (1.7) can be taken to include $n = 0$. Also, since the exponential distribution has no memory, the same result would follow for any interval of duration t. In virtue of this result, the term 'Poisson stream of events with rate λ' is equivalent to a completely random stream of events. Some basic properties of the Poisson distribution are reviewed in the following problems. In particular, from problem 5(i), that λ is the mean number of events per unit time.

Problem 5. A discrete, non-negative random variable X with p.d.f.

$$f(x) = \frac{\lambda^x \exp(-\lambda)}{x!} \,, x = 0,1, \ldots$$

is said to have the Poisson distribution with parameter λ.
Show that
(i) $E(X) = \lambda$, $V(X) = \lambda$
(ii)   the m.g.f. of X is $\exp (-\lambda) \exp [\lambda \exp (\theta)]$. ∎

Problem 6. X_i $(i = 1,2, \ldots, n)$ is a sequence of independent random variables. The ith random variable has a Poisson distribution with parameter λ_i.
(i)   Show that $\sum_1^n X_i$ has a Poisson distribution with parameter $\sum_1^n \lambda_i$.

(ii)   If $p_i = \lambda_i / \sum_1^n \lambda_i$, show that the conditional probability that $X_i = x_i$ given

$$\sum_1^n X_j = m \text{ is } \binom{m}{x_i} p_i^{x_i} (1-p_i)^{m-x_i}$$

As an application, consider n simultaneous and independent Poisson streams with rates λ_i. Then (ii) provides the conditional probability that there are x_i events in stream i, given that there are m events in all. ∎

Further applications of the Poisson distribution

Example 1
Suppose that the number of events in the interval $(0,t)$ has a Poisson distribution with parameter λt but each event has only a probability p of being recorded. It is required to find the distribution of the number of *recorded* events.

$$Pr\,[n \text{ recorded}] = \sum_{m=n}^{\infty} Pr(n \text{ recorded} \mid m \text{ events})\,Pr(m \text{ events})$$

$$= \sum_{m=n}^{\infty} \binom{m}{n} p^n (1-p)^{m-n} (\lambda t)^m \exp(-\lambda t)/m!$$

$$= \exp(-\lambda t)\frac{(p\lambda t)^n}{n!} \sum_{m=n}^{\infty} \frac{[(1-p)\lambda t]^{m-n}}{(m-n)!}$$

$$= \frac{(p\lambda t)^n \exp(-p\lambda t)}{n!}.$$

The distribution of the number of recorded events is again Poisson but with parameter $p\lambda t$.

Example 2
Suppose the number of insurance claims arising in a fixed period has a Poisson distribution with parameter λ. The claims are for independent amounts with a common distribution. Let $M(\theta)$ be the m.g.f. of this common distribution, then it is possible to find the moment generating function for the total amount claimed. Suppose there are n claims and the amount for the ith is X_i, then

$$E\left[\exp\left(\sum_{1}^{n} X_i \theta\right) \mid n\right] = [E\{\exp(X\theta)\}]^n = [M(\theta)]^n.$$

Hence the unconditional expectation is the mean of $[M(\theta)]^n$ over the distribution of the number of claims, viz.

$$\sum_{n=0}^{\infty} [M(\theta)]^n \, Pr\,(N=n)$$

$$= \sum_{0}^{\infty} [M(\theta)]^n \, \lambda^n \exp(-\lambda)/n!$$

$$= \exp[-\lambda+\lambda M(\theta)].$$

The reader should confirm that for the case $X_i = 1$ with probability p and $X_i = 0$ otherwise, we recover the result in Example 1.

Example 3

A possible model for factory accidents is that each member of the work force experiences accidents at a Poisson rate, but the mean number is particular to each member. We examine how the distribution of accidents for a member chosen at random may be derived for one such model. The number of accidents, X, has a Poisson distribution with parameter θ but θ is a value of a random variable with, for example, a gamma distribution with parameters (n,λ). It is required to find the unconditional distribution of X.

$$Pr[X = x \mid \theta] = \frac{\theta^x \exp(-\theta)}{x!}$$

$$Pr(X = \text{x}) = \int_0^\infty Pr[X = x \mid \theta] \, f(\theta) \, d\theta$$

$$= \int_0^\infty \frac{\theta^x e^{-\theta}}{x!} \cdot \frac{\lambda(\lambda\theta)^{n-1} e^{-\lambda\theta}}{(n-1)!} \, d\theta$$

$$= \frac{\lambda^n}{x! \, (n-1)!} \int_0^\infty \theta^{n+x-1} e^{-\theta \, (\lambda+1)} \, d\theta$$

$$= \frac{(n+x-1)!}{x! \, (n-1)!} \frac{\lambda^n}{(1+\lambda)^{n+x}}.$$

X has a negative binomial distribution for which the variance exceeds the mean. For the Poisson distribution these moments are equal — worth bearing in mine for data of this type.

1.3 THE POISSON PROCESS

We now consider more closely another implication of the assumption that the time between events has an exponential distribution. We ask, what is the probability that an event occurs between t and $t+\delta t$, where δt is small? We say 'an event' since it is not necessarily the first. However, since the exponential distribution has no memory, this is entirely equivalent to asking 'at time t, what is the probability of the *next* event falling in the interval $(t,t+\delta t)$?' This is, from (1.3), $1-\exp(-\lambda\delta t) = \lambda\delta t-(\lambda\delta t)^2/2!+\ldots = \lambda\delta t+o(\delta t)$. *This probability includes some contribution from two or more events but we shall now show that this contribution is $o(\delta t)$. At time t, regardless of the time since the last event, the additional time, y, to the next event has an exponential distribution with parameter λ as does the further time to the event immediately following. Hence the probability of two or more events in $(t,t+\delta t)$ is

$$\int_0^{\delta t} \lambda \exp(-\lambda y) \{1-\exp[-\lambda(\delta t-y)]\} \, dy \quad = 1-(1+\lambda\delta t)\exp(-\lambda\delta t)$$

$$< 1-(1+\lambda\delta t)(1-\lambda\delta t)$$

$$= (\lambda\delta t)^2 = o(\delta t).$$

The foregoing may be crudely summarised by stating that the probability of just one event in $(t,t+\delta t)$ is $\lambda\delta t$ and of more than one event is negligible. We will now show conversely that if this is the case, the stream of events must be Poisson.

Suppose events are happening in time so that, independently of the number of events that have already occurred in $(0,t)$,

(i)   the probability of just one event in the interval $(t,t+\delta t)$ is $\lambda\delta t+o(\delta t)$,
(ii) the probability of more than one event in $(t,t+\delta t)$ is $o(\delta t)$, and hence
(iii) the probability of no event in the interval $(t,t+\delta t)$ is $1-\lambda\delta t+o(\delta t)$,

then we may show that the number of events in $(0,t)$ has a Poisson distribution with parameter λt.

(Note: the various terms which are $o(\delta t)$ are not the same and must sum to zero to conserve a total probability of one!)

Let $p_n(t)$ be the probability that just n events happen in $(0,t)$. We now consider the extended interval $(0,t+\delta t)$. Then there will be just $n(\geqslant 1)$ events by $t+\delta t$ if there had been either n events in $(0,t)$ and no event in $(t,t+\delta t)$ or fewer than n events in $(0,t)$ and the number was made up in $(t,t+\delta t)$. The first possibility has probability $p_n(t)[1-\lambda\delta t+o(\delta t)]$. For the second possibility, $(n-1)$ events by time t followed by just one event in $(t,t+\delta t)$ has probability $p_{n-1}(t)[\lambda\delta t+o(\delta t)]$ and other cases have a total probability which is $o(\delta t)$. [Notice that we have used the independence of events in the intervals $(0,t)$, $(t,t+\delta t)$.] Hence

$$p_n(t+\delta t) = p_n(t)[1-\lambda\delta t+o(\delta t)] + p_{n-1}(t)[\lambda\delta t+o(\delta t)] + o(\delta t), \tag{1.8}$$

$$\frac{p_n(t+\delta t)-p_n(t)}{\delta t} = -\lambda p_n(t) + \lambda p_{n-1}(t) + \frac{o(\delta t)}{\delta t}.$$

Now let $\delta t \to 0$ and we have, in the limit,

$$\frac{dp_n(t)}{dt} = \lambda p_{n-1}(t) - \lambda p_n(t), n \geqslant 1. \tag{1.9}$$

*   We say that $Q(\delta t)$ is $o(\delta t)$ if $\lim_{\delta t\to 0}[Q(\delta t)/\delta t] = 0$, that is to say, $Q(\delta t)$ has a smaller order of magnitude than δt.

The case $n = 0$ appears superficially to be different, for the only possibility is that there was no event in $(0,t)$ *and* subsequently no event in $(t,t+\delta t)$. Thus

$$p_0(t+\delta t) = p_0(t)\,[1-\lambda\delta t+o(\delta t)],$$

$$\frac{p_0(t+\delta t)-p_0(t)}{\delta t} = -\lambda p_0(t) + \frac{o(\delta t)}{\delta t},$$

$$\frac{dp_0(t)}{dt} = -\lambda p_0(t). \tag{1.10}$$

In fact, (1.10) is (1.9) for $n = 0$ if we set $p_{-1}(t) \equiv 0$. We can easily show that $p_n(t) = (\lambda t)^n \exp(-\lambda t)/n!$ satisfies (1.9) by the method of induction. For this, we need a starting point, the easiest being provided by the case $n = 0$, which can be solved explicitly.

We have

$$\frac{p_0'(t)}{p_0(t)} = -\lambda,$$

$$\frac{d}{dt}\,[\log\{p_0(t)\}\,] = -\lambda$$

$$\log\{p_0(t)\} = -\lambda t + c,$$

$$p_0(t) = \exp(-\lambda t + c). \tag{1.11}$$

But when $t = 0, p_0(0) = 1$, thus the integration constant is zero. Hence, $p_0(t) = \exp(-\lambda t)$ is of the required form. Now assume that $p_{n-1}(t) = (\lambda t)^{n-1}\exp(-\lambda t)/(n-1)!$ and substituting in (1.9),

$$\frac{dp_n(t)}{dt} + \lambda p_n(t) = \frac{\lambda(\lambda t)^{n-1}\exp(-\lambda t)}{(n-1)!}.$$

Multiplying by the integrating factor, $\exp(\lambda t)$,

$$\frac{dp_n(t)}{dt}\exp(\lambda t) + p_n(t)\,\lambda\exp(\lambda t) = \frac{\lambda^n t^{n-1}}{(n-1)!},$$

$$\frac{d}{dt}\,[p_n(t)\exp(\lambda t)] = \frac{\lambda^n t^{n-1}}{(n-1)!},$$

$$p_n(t)\exp(\lambda t) = \frac{\lambda^n t^n}{n!} + c_n.$$

Since, $p_n(0) = 0(n > 0)$, $c_n = 0$ and therefore
$$p_n(t) = (\lambda t)^n \exp(-\lambda t)/n! \tag{1.12}$$

The usual induction argument then shows that formula (1.12) is in fact correct, and is the p.d.f. of a Poisson distribution with parameter λt.

Problem 7. The time to the nth event is less than t if and only if there are at least n events in the interval $(0,t)$. Hence if $f_n(.)$ is the p.d.f. of the time to the nth event, show that

$$\int_0^t f_n(x)\mathrm{d}x = \sum_{k=n}^{\infty} p_k(t)$$

where $p_k(t)$ is as defined in equation (1.12). Differentiate both sides with respect to t to show that the time to the nth event has a gamma distribution with parameters (n,λ). ∎

Problem 8. By multiplying equation (1.9) by θ^n and summing over n, show that $G(\theta,t) = \sum_0^{\infty} p_n(t)\,\theta^n$ satisfies

$$\frac{\partial G}{\partial t} = \lambda\,(\theta-1)\,G.$$

Solve for G and, by expanding as a power series, recover $p_n(t)$.

Problem 9. If in the postulates we everywhere replace λ by $\lambda(t)$, prove that the number of events in $(0,t)$ has a Poisson distribution with parameter $\int_0^t \lambda(u)\mathrm{d}u$.
Hence, or otherwise, prove that if $\lambda(u) = 1/(1+\alpha u)$, the p.d.f. of the time to the first event is $f(t) = 1/[1+\alpha t]^{(1+\alpha)/\alpha}$. ∎

1.4 THE TWO-STATE PROCESS

Suppose on a single telephone line, calls arrive in a Poisson stream at the rate λ per unit time. If a caller finds the line engaged then his call is lost. If a caller finds the line free then a conversation begins, the duration of which has an exponential distribution with mean $1/\mu$. Thus the line is in one of two states — free or busy — and the changes between them have independent waiting times which have exponential distributions with parameters λ,μ respectively. We can compute the probabilities that the line is free (state A) or busy (state B) at time t, and these will depend on the initial state. More precisely, let $p_{AA}(t)$ be the probability that, given the line is initially free, it is again free at a time t later. Let $p_{AB}(t)$ be the probability that the line is busy at time t given that it is initially free. Now the line is free at time $t+\delta t$ if either it was free at time t and remained so for a further time δt, or it was busy at time t and became free in the further time δt. In terms of the corresponding probabilities,

$$p_{AA}(t+\delta t) = [1-\lambda\delta t+o(\delta t)]\, p_{AA}(t) + [\mu\delta t+o(\delta t)]\, p_{AB}(t) + o(\delta t)$$

$$\frac{p_{AA}(t+\delta t)-p_{AA}(t)}{\delta t} = -\lambda p_{AA}(t)+\mu p_{AB}(t) + \frac{o(\delta t)}{\delta t}.$$

Now let $\delta t \to 0$, and, substituting for $p_{AB}(t) = 1-p_{AA}(t)$, we have

$$p'_{AA}(t) = -(\lambda+\mu)p_{AA}(t)+\mu.$$

This differential equation is readily solved after multiplying by the integrating factor $\exp[(\lambda+\mu)t]$, when we have

$$\frac{d}{dt}\left[\exp\left\{(\lambda+\mu)t\right\} p_{AA}(t)\right] = \mu \exp[(\lambda+\mu)t],$$

$$\exp[(\lambda+\mu)t]p_{AA}(t) = \frac{\mu}{\lambda+\mu}\exp[(\lambda+\mu)t] + c,$$

where c is the constant of integration. But $p_{AA}(0) = 1$, hence $c = \lambda/(\lambda+\mu)$, and

$$p_{AA}(t) = \frac{\mu}{\lambda+\mu} + \frac{\lambda}{\lambda+\mu}\exp[-(\lambda+\mu)t],$$

$$p_{AB}(t) = \frac{\lambda}{\lambda+\mu} - \frac{\lambda}{\lambda+\mu}\exp[-(\lambda+\mu)t].$$

By interchanging λ and μ we immediately obtain

$$p_{BB}(t) = \frac{\lambda}{\lambda+\mu} + \frac{\mu}{\lambda+\mu}\exp[-(\lambda+\mu)t],$$

$$p_{BA}(t) = \frac{\mu}{\lambda+\mu} - \frac{\mu}{\lambda+\mu}\exp[-(\lambda+\mu)t].$$

The limiting values of these probabilities are interesting.

$$\lim_{t\to\infty}[p_{AA}(t)] = \lim_{t\to\infty}[p_{BA}(t)] = \mu/(\lambda+\mu),$$

$$\lim_{t\to\infty}[p_{AB}(t)] = \lim_{t\to\infty}[p_{BB}(t)] = \lambda/(\lambda+\mu).$$

That is, after a long time, the probabilities of being free or busy do not depend on the initial state. Moreover, if we fancifully suppose that the initial state is decided by a chance mechanism so that it is free with probability $\mu/(\lambda+\mu)$ and busy with probability $\lambda/(\lambda+\mu)$ when the *unconditional* probability of the line being free at *any* later time t is

$$\frac{\mu}{\lambda+\mu}p_{AA}(t) + \frac{\lambda}{\lambda+\mu}p_{BA}(t) = \frac{\mu}{\lambda+\mu}.$$

In the sense just discussed, $\mu/(\lambda+\mu)$, $\lambda/(\lambda+\mu)$ are said to be *stationary* probabilities for the two-state system. ∎

Problem 10. Consider the following alternative derivation for the probabilities of the two-state system. If initially the system is in state A, it will again be in state A at time t if either

(1)   in the interval $(0,\delta t)$ it remains in state A and is subsequently in A after a duration $t-\delta t$, or

(2)   in the interval $(0,\delta t)$ it switches to state B and is subsequently in state A after a duration $t-\delta t$.

Hence show that

$$p_{AA}'(t) = \lambda p_{BA}(t) - \lambda p_{AA}(t)$$

and similarly

$$p_{BA}'(t) = \mu p_{AA}(t) - \mu p_{BA}(t).$$

By differentiating the first of these equations and substituting for $p_{BA}'(t)$, show that

$$p_{AA}''(t) = -(\lambda+\mu)p_{AA}'(t)$$

and solve for $p_{AA}(t)$. ∎

Problem 11. For the two-state system, if initially in state A, the system will be in A at time t if either

(1)   the system remains continuously in A for at least time t, or

(2)   the system switches to B at time x and is in state A after a further time $t-x$ $(0<x<t)$.

Hence, for switching times exponentially distributed, show that

$$p_{AA}(t) = \exp(-\lambda t) + \int_0^t \lambda \exp(-\lambda x)p_{BA}(t-x)\,dx$$

$$= \exp(-\lambda t) + \exp(-\lambda t)\int_0^t \lambda \exp(\lambda y)p_{BA}(y)\,dy.$$

Deduce that

$$p_{AA}'(t) = -\lambda p_{AA}(t) + \lambda p_{BA}(t)$$

and that

$$p_{AA}''(t) = -(\lambda+\mu)p_{AA}'(t).$$ ∎

Problem 12. On a single telephone line, calls arrive in a Poisson stream with intensity λ_1. Any particular caller either finds the line busy, in which case his call is 'ineffective' or finds it free, in which case he makes connection and his call is 'effective'. The effective calls have durations which are independent and have a common exponential distribution with parameter λ_2. Calculate:

(a)   the probability that exactly k callers find the line busy in the course of one connected call;

(b)   the expectation of the ratio of busy to non-busy time at the end of n effective calls, starting at an arbitrary time when the line is free. ■

Expected time in a State

In a sequence of n independent trials with probability of 'success' p on each trial, the proportion of successes has expected value p. However, in the two-state process, with exponential switching times, the probability of being in a particular state depends on the time so we cannot recover such a simple relationship for the proportion of time spent in a state. Let $T_{AA}(t)$ be the (random) time spent in state A, given that the process was initially in state A, in the course of the fixed interval $(0,t)$. There are several rather elegant ways of computing $E[T_{AA}(t)]$ — see Cox and Miller [1] — but we shall employ a less sophisticated approach. Now it is either the case that the system stays in state A for at least time t, with probability $\exp(-\lambda t)$, or it switches to B at $x(< t)$ and then spends some further time in A in the interval (x,t). In terms of expectations,

$$E[T_{AA}(t)] = t \exp(-\lambda t) + \int_0^t \left\{ x + E[T_{BA}(t-x)] \right\} \lambda \exp(-\lambda x) dx, \quad (1.13)$$

where $T_{BA}(t)$ is the time spent in state A given that the process was initially in state B in the course of a fixed interval $(0,t)$.

$$E[T_{AA}(t)] = t \exp(-\lambda t) +$$

$$\int_0^t x. \lambda \exp(-\lambda x) dx + \exp(-\lambda t) \int_0^t \lambda \exp(\lambda y) E[T_{BA}(y)] dy. \quad (1.14)$$

After differentiating with respect to t,

$$E'[T_{AA}(t)] = \exp(-\lambda t) - \lambda \exp(-\lambda t) \int_0^t \lambda \exp(\lambda y) E[T_{BA}(y)] dy +$$

$$\lambda E[T_{BA}(t)],$$

which, after substituting for the integral from equation (1,14), reduces to

$$E'[T_{AA}(t)] = 1 + \lambda E[T_{BA}(t)] - \lambda E[T_{AA}(t)].$$

But if initially the system is in state B, then either it spends no time in A, with probability exp $(-\mu t)$, or it switches to A at some intermediate time x and subsequently spends some time in A in the remaining duration $t-x$. That is,

$$E\,[T_{BA}\,(t)] = \int_0^t \mu \exp(-\mu x)\,E[T_{AA}(t-x)]\,dx$$

$$= \exp(-\mu t) \int_0^t E[T_{AA}(y)]\,\mu \exp(\mu y)dy.$$

The reader is now in a position to confirm, as a problem, that

$$E''\,[T_{AA}(t)] = \mu - (\lambda+\mu)E'\,[T_{AA}(t)].$$

This is a comparatively easy differential equation to solve and yields

$$E[T_{AA}(t)] = \frac{\mu t}{\lambda+\mu} + \frac{\lambda}{(\lambda+\mu)^2}\,\{1-\exp[-(\lambda+\mu)t]\}.$$

It is to be remarked that

$$\lim_{t\to\infty}\,[E[T_{AA}(t)]/t] = \frac{\mu}{\lambda+\mu},$$

or the expected proportion of time spent in state A tends to the stationary probability of being in state A.

Problem 13. Let $p_n(t\,|A)$ be the probability that just n calls are connected in the interval $(0,t)$, given that the line was initially free. Show for $n \geqslant 2$,

$$p_n(t\,|A) = \exp(-\lambda t) \int_0^t \lambda \exp(\lambda y)p_n(y\,|B)dy,$$

$$p_n(t|B) = \exp(-\mu t)\int_0^t \mu \exp(\mu y)p_{n-1}(y|A)dy,$$

where $p_n(t\,|B)$ is the probability that just n calls are connected, given that the line was initially busy. Hence,

$$p_n'(t|A) = -\lambda p_n(t|A) + \lambda p_n(t|B).$$
$$p_n'(t|B) = -\mu p_n(t|B) + \mu p_{n-1}(t|A),$$
$$p_n''(t|A) = -(\lambda+\mu)p_n'(t|A) - \lambda\mu[p_n(t|A) - p_{n-1}(t|A)].$$

Note: if an interval starts with a call in progress, then it is to be counted as one of the n.

REFERENCE

[1] *The Theory of Stochastic Processes*, D. R. Cox and H. D. Miller. Methuen, London, 1965.

BRIEF SOLUTIONS AND COMMENTS ON THE PROBLEMS

Problem 1

$$\int_0^\infty x\lambda \exp(-\lambda x)dx = \left[-x \exp(-\lambda x)\right]_0^\infty + \int_0^\infty \exp(-\lambda x)dx$$

$$= 0 - \left[\frac{\exp(-\lambda x)}{\lambda}\right]_0^\infty = \frac{1}{\lambda}.$$

$$\frac{d}{d\theta}\left(\frac{\lambda}{\lambda-\theta}\right) = \frac{\lambda}{(\lambda-\theta)^2}. \text{ Value at } \theta = 0 \text{ is } \frac{1}{\lambda}.$$

Problem 2.   When $i-1$ have failed, the extra time to the next failure is the minimum of $k-(i-1)$ exponentially distributed random variables.

$$\sum_1^k Z_i = T_k = \max [X_1, X_2, \ldots, X_k]. \text{ Hence}$$

$$Pr(T_k < t) = \Pi \ Pr(X_i < t) = [1 - \exp(-\lambda t)]^k.$$

Problem 3. Integrate by parts.

(i) $\int_0^\infty x^{a-1} \exp(-x)dx = \left[-x^{\alpha-1} \exp(-x)\right]_0^\infty + (\alpha-1) \int_0^\infty x^{\alpha-2} \exp(-x)dx.$

if $\alpha > 1, \lim_{x \to 0} (x^{\alpha-1}) = 0$, hence result. $\Gamma(m) = (m-1)\Gamma(m-1)$

$= (m-1)!\Gamma(1) = (m-1)!$, when m is a positive integer.

(ii) Change variable, set $x = y^2$. Then $y = z/\sqrt{2}, \alpha = \frac{1}{2}$ and note that for the standard normal distribution

$$\int_{-\infty}^\infty \frac{1}{\sqrt{2\pi}} \exp(-z^2/2) \ dz = 1.$$

Problem 4

(i) $E(X) = \int_0^\infty \frac{\lambda^\alpha x^\alpha \exp(-\lambda x)}{\Gamma(\alpha)} \ dx = \frac{\Gamma(\alpha+1)}{\lambda\Gamma(\alpha)} \int_0^\infty \frac{\lambda(\lambda x)^{(\alpha+1)-1} \exp(-\lambda x)}{\Gamma(\alpha+1)} \ dx$

$$= \frac{\Gamma(\alpha+1)}{\lambda\Gamma(\alpha)} \cdot 1 = \frac{\alpha}{\lambda}, \text{ using problem 3(i).}$$

Similarly $E(X^2) = \alpha\,(\alpha+1)/\lambda^2$ and $V(X) = E(X^2) - E^2(X)$.

$$E[\exp(X\theta)] = \int_0^\infty \frac{\lambda(\lambda x)^{\alpha-1}\exp[-x(\lambda-\theta)]}{\Gamma(\alpha)}\,dx. \text{ Now put } x(\lambda-\theta) = y.$$

(ii) $$E(1/X) = \int_0^\infty \frac{\lambda^\alpha x^{\alpha-2}\exp(-\lambda x)}{\Gamma(\alpha)}\,dx = \frac{\lambda\Gamma(\alpha-1)}{\Gamma(\alpha)}\int_0^\infty \frac{\lambda(\lambda x)^{\alpha-2}\exp(-\lambda x)}{\Gamma(\alpha-1)}\,dx$$

$$= \frac{\lambda\Gamma(\alpha-1)}{\Gamma(\alpha)} = \frac{\lambda}{\alpha-1}, \text{ if } \alpha > 1.$$

$\log f(x) = \alpha\log\lambda + (\alpha-1)\log x - \lambda x - \log\Gamma(\alpha)$.

$\dfrac{d}{dx}[\log f(x)] = \dfrac{\alpha-1}{x} - \lambda$. Maximum at $x = (\alpha-1)/\lambda$.

(iii) The moment generating of X_i is $[\lambda/(\lambda-\theta)]^{\alpha_i}$.

Since the X_i are independent, the moment generating function of $\sum_1^n X_i$, is

$[\lambda/(\lambda-\theta)]^{\sum_1^n \alpha_i}$. But this is the moment generating function of another

gamma distribution with parameters $\left(\sum_1^n \alpha_i, \lambda\right)$.

Problem 5

(i) $E(X) = \sum_0^\infty \dfrac{x.\lambda^x\exp(-\lambda)}{x!} = \lambda\exp(-\lambda)\sum_1^\infty \dfrac{\lambda^{x-1}}{(x-1)!} = \lambda$.

Similarly, $E[X(X-1)] = \lambda^2$. $V(X) = E[X(X-1)] + E(X) - E^2(X) = \lambda^2 + \lambda - \lambda^2 = \lambda$.

(ii) $M_X(\theta) = E[\exp(X\theta)] = \sum_{x=0}^\infty \dfrac{\exp(x\theta)\lambda^x\exp(-\lambda)}{x!}$

$$= \exp(-\lambda)\sum_{x=0}^\infty \frac{[\lambda\exp(\theta)]^x}{x!} = \exp(-\lambda)\exp[\lambda\exp(\theta)].$$

Problem 6

(i) The m.g.f. of X_i is $\exp(-\lambda_i)\exp[\lambda_i\exp(\theta)]$. Since the X_i are independent,

the m.g.f. of $\sum_1^n X_i$ is $\exp\left(-\sum_1^n \lambda_i\right)\exp\left[\sum_1^n \lambda_i\exp(\theta)\right]$. But this is the

m.g.f. of another Poisson distribution having parameter $\sum_1^n \lambda_i$.

(ii) $Pr[X_i = x_i \mid \sum_1^n X_j = m] = Pr[X_i = x_i \text{ and } \sum_1^n X_j = m] / Pr\left[\sum_1^n X_j = m\right]$

$= Pr[X_i = x_i] \, Pr\left[\sum_{j \neq i}^n X_j = m - x_i\right] / Pr\left[\sum_1^n X_j = m\right].$

$\sum_1^n X_j$ has a Poisson distribution with parameter $\sum_1^n \lambda_i$. $\sum_{j \neq i}^n X_j$ has a Poisson

distribution with parameter $\sum_1^n \lambda_j - \lambda_i$, X_i has a Poisson distribution with

parameter λ_i. Substitute the appropriate p.d.f.s to obtain result.

Problem 7

$$\sum_{k=n}^{\infty} \frac{d}{dt}\left[\frac{(\lambda t)^k \exp(-\lambda t)}{k!}\right] = \sum_{k=n}^{\infty} \left[\frac{k\lambda(\lambda t)^{k-1}}{k!} - \frac{\lambda(\lambda t)^k}{k!}\right] \exp(-\lambda t)$$

$$= \lambda \exp(-\lambda t) \sum_{k=n}^{\infty} \left[\frac{(\lambda t)^{k-1}}{(k-1)!} - \frac{(\lambda t)^k}{k!}\right] = \frac{\lambda(\lambda t)^{n-1} \exp(-\lambda t)}{(n-1)!}$$

which is the p.d.f. of a gamma distribution with parameters (n, λ).

Problem 8

$$\sum_{n=0}^{\infty} \theta^n \frac{dp_n(t)}{dt} = \frac{\partial}{\partial t}\left[\sum_{n=0}^{\infty} \theta^n p_n(t)\right] = \frac{\partial G}{\partial t}. \quad \sum_{n=0}^{\infty} \lambda \theta^n p_n(t) = \lambda G,$$

$$\sum_{n=1}^{\infty} \lambda \theta^n p_{n-1}(t) = \lambda \theta \sum_1^{\infty} \theta^{n-1} p_{n-1}(t) = \lambda \theta G, \text{ hence result.}$$

The equation may be written, $\frac{\partial}{\partial t}(\log G) = \lambda(\theta - 1)$. Hence $\log G = \lambda(\theta - 1)t + k(\theta)$. But $G(\theta, 0) = 1$, hence $k(\theta) = 0$.

$G = \exp[\lambda(\theta - 1)t] = \exp(-\lambda t) \sum_{n=0}^{\infty} \frac{(\lambda t \theta)^n}{n!}$ and the coefficient of θ^n is $(\lambda t)^n \exp(-\lambda t)/n!$.

Problem 9. Using the kind of argument which led to equation (1.8),

$$p_n(t + \delta t) = p_n(t)[1 - \lambda(t)\delta t] + p_{n-1}(t)\lambda(t)\delta t + o(\delta t).$$

$$p_n'(t) = -\lambda(t)p_n(t) + \lambda(t)p_{n-1}(t).$$

For convenience, write $\phi(t) = \int_0^t \lambda(u)du$. We write the last equation as

$$p_n'(t) + \phi'(t)p_n(t) = \phi'(t)p_{n-1}(t).$$

Assume the result for $n-1$, multiply by the integrating factor $\exp[+\phi(t)]$ and verify that $p_n(t)$ is also of the required form. The p.d.f. of the time to the first event is $p_0(t) \lambda(t)$, and $p_0(t) = \exp[-\phi(t)]$ where $\phi(t) = \int_0^t \dfrac{du}{1+\alpha u} = \log(1+\alpha t)^{1/\alpha}$.

Problem 10

$$P_{AA}(t) = (1-\lambda\delta t)p_{AA}(t-\delta t) + \lambda\delta t p_{BA}(t-\delta t)$$

$$\Rightarrow \quad P_{AA}'(t) = \lambda p_{BA}(t) - \lambda p_{AA}(t).$$

$$P_{BA}(t) = \mu\delta t p_{AA}(t-\delta t) + (1-\mu\delta t)p_{BA}(t-\delta t)$$

$$\Rightarrow \quad p_{BA}'(t) = \mu p_{AA}(t) - \mu p_{BA}(t).$$

To obtain the next result, differentiate the first equation, substitute for $p_{BA}'(t)$ and substitute for $p_{BA}(t)$ from the first equation. We have

$$\frac{d}{dt}[\log\{p_{AA}'(t)\}] = -(\lambda+\mu),$$

$$p_{AA}'(t) = k \exp[-(\lambda+\mu)t].$$

Now from the first equation, $p_{AA}'(0) = +\lambda p_{BA}(0) - \lambda p_{AA}(0) = -\lambda$. After integrating once more.

$$P_{AA}(t) = \frac{\mu}{\lambda+\mu} + \frac{\lambda}{\lambda+\mu} \exp[-(\lambda+\mu)t].$$

Problem 11. The system remains in A with probability $\exp(-\lambda t)$. The time to the switch to B has p.d.f. $\lambda \exp(-\lambda x)$, and the probability of being in A again after a further time $t-x$ is $p_{BA}(t-x)$. Hence total probability of the event in (2) is $\int_0^t \lambda \exp(-\lambda x)p_{BA}(t-x)dx$. Now set $y = t-x$ and then differentiate $p_{AA}(t)$ with respect to t, to obtain $p_{AA}'(t) = -\lambda p_{AA}(t) + \lambda p_{BA}(t)$. A similar argument yields

$$p_{BA}(t) = \int_0^t \mu \exp(-\mu x)p_{AA}(t-x)dx$$

$$= \exp(-\mu t) \int_0^t \mu \exp(\mu y)p_{AA}(y)dy.$$

Problem 12

(a) If the call has length t, the number of incoming calls has a Poisson distribution with parameter $\lambda_1 t$. $Pr[k|t] = (\lambda_1 t)^k \exp(-\lambda_1 t)/k!$. Since duration of call has exponential distribution, the unconditional probability of k calls is

$$\int_0^\infty \frac{(\lambda_1 t)^k \exp(-\lambda_1 t) \, \lambda_2 \exp(-\lambda_2 t)}{k!} \, dt,$$

$$= \frac{\lambda_1^k \lambda_2}{k!} \int_0^\infty t^k \exp\left[-(\lambda_1 + \lambda_2)t\right] \, dt = \frac{\lambda_1^k \lambda_2}{(\lambda_1 + \lambda_2)^{k+1}}$$

(b) Total busy time, B, has a $\Gamma(n, \lambda_2)$ distribution. Total free time, F, has a $\Gamma(n, \lambda_1)$ distribution. $E[B/F] = E(B)E(1/F) = n\lambda_1/[(n-1)\lambda_2]$.

Problem 13. The time to the first call has p.d.f. $\lambda \exp(-\lambda x)$. The probability that there are n connections in the remaining time $t-x$ is $p_n(t-x|B)$. Hence

$$p_n(t|A) = \int_0^t \lambda \exp(-\lambda x) p_n(t-x|B) dx.$$

Substitute $t - x = y$.

$$p_n'(t|A) = -\lambda \exp(-\lambda t) \int_0^t \lambda \exp(\lambda y) p_n(y|B) dy +$$

$$\exp(-\lambda t) \lambda \exp(\lambda t) p_n(t|B)$$

$$= -\lambda p_n(t|A) + \lambda p_n(t|B).$$

A similar argument gives $p_n'(t|B)$. Differentiate $p_n'(t|A)$ again to obtain

$$p_n''(t|A) = -\lambda p_n'(t|A) + \lambda p_n'(t|B).$$

Substitute for $p_n'(t|B)$, and then for $p_n(t|B)$ from the expression for $p_n'(t|A)$.

CHAPTER 2

Birth and Death Processes

2.1 INTRODUCTION

It is likely enough that the reader first met a growth process in connection with
the mundane topic of compound interest. Under this regime, interest is added to
the principal at regular intervals and the *total* earns interest subsequently.
Consider investing an initial amount A for a time t at an interest rate which is a
proportion R of the accumulated investment per unit time. Divide t into n
intervals of duration t/n and suppose interest is added at the end of each such
interval. If A accumulates to $A(t)$ at time t, then

$$\begin{aligned}
A(t) &= A(t-t/n) + (Rt/n)A(t-t/n) \\
&= (1+Rt/n)A(t-t/n) \\
&= (1+Rt/n)^n A(0) = (1+Rt/n)^n A,
\end{aligned}$$

after repeated application. This formula is useful for comparing the effect of
different values of R,n for the same time t. A good approximation, for large n, is
obtained by noting that

$$A \lim_{n \to \infty} (1+Rt/n)^n = A \exp(Rt). \tag{2.1}$$

This (limiting) amount may also be regarded as the accumulated capital if
interest were added 'instantaneously'.

We may obtain (2.1) directly if we suppose that the rate of growth of $A(t)$ is
proportional to $A(t)$. That is,

$$\frac{d[A(t)]}{dt} = RA(t),$$

or

$$\frac{d}{dt} [\log A(t)] = R,$$

which implies

$$A(t) = A \exp(Rt).$$

We call this a *deterministic growth process* if $R > 0$. It is deterministic in the sense that the final amount is completely determined by the rate of growth and the duration of the time in which growth takes place. (For a discussion of deterministic growth, decay and related models see Burghes and Wood [1].)

Our interest, however, will centre on growth and decay processes for which the instants of increase or decrease in the size of the 'population' are not fixed but have probability distributions. The easiest example is the **pure linear birth process**. In this, the population consists of a number of identical organisms. Independently of each other, each organism gives birth to just one similar organism. The times between births have common exponential distributions. We can now only attempt to calculate the probabilities of certain events. For example, if initially there is one such organism present, what is the probability that there are just n present at time t later? We first give a direct argument.

Suppose that, at any time, there are r organisms present. The individual times for each of these to produce another organism is exponentially distributed, say with parameter λ. Hence the time to the next birth is the minimum of r such independent birth times and hence has an exponential distribution with parameter $r\lambda$ (see Problem 2, Chapter 1). Thus, the total time, T_n, to the nth birth has the form $T_n = \sum_{i=1}^{n} Y_i$, where Y_i has the exponential distribution with parameter $i\lambda$. At the moment that the $(n-1)$th birth takes place, the number of organisms present is n, since one was present initially. Hence, if N_t is the size of the population by time t,

$$N_t \geqslant n \leftrightarrow T_{n-1} \leqslant t.$$

Hence,
$$
\begin{aligned}
Pr[N_t = n] &= Pr[N_t \geqslant n] - Pr[N_t \geqslant n+1], \\
&= Pr[T_{n-1} \leqslant t] - Pr[T_n \leqslant t] \\
&= (1-e^{-\lambda t})^{n-1} - (1-e^{-\lambda t})^n, \text{ from Problem 2 Chapter 1,} \\
&= (1-e^{-\lambda t})^{n-1} e^{-\lambda t}.
\end{aligned}
\tag{2.2}
$$

From (2.2), we note that N_t has a distribution with mean $\exp(\lambda t)$. This agrees with (2.1) when $A = 1$, $R = \lambda$. Thus this probabilistic model agrees with the deterministic model in expectation.

In spite of its limitations, this pure birth process has been used in important applications. In Yule's study of the creation of new species, the size of the population is the number of species and birth consists of one new species by mutation [2]. The same model was used by Furry in connection with the emission of cosmic rays [3]. Here, the size of the population is the number of particles present and birth consists of a particle splitting into replicas. Indeed, the pure birth process is also known as the Yule–Furry process. We next propose to re-examine this same process from several other points of view. This is in order to accumulate a variety of techniques which may also be applied to processes of allied types. The reader is warned that it is seldom possible to find direct methods similar to that used to obtain (2.2). With the same end in view, we present the model in a framework which more readily admits further generalisation.

2.2 THE PURE BIRTH PROCESS

A colony of organisms begins with one member. At any later time the probability that it splits into two during a small interval of duration δt is $\lambda \delta t + o(\delta t)$ and the probability that it does not split at all is $1 - \lambda \delta t + o(\delta t)$. All descendants of the original member independently follow the same pattern. We seek the probability that at any time there are just $n (\geqslant 1)$ members in the colony. colony.

Let $p_{ij}(t)$ be the probability that in any interval of duration t, the colony grows from i to j members. We can decompose the event n members at time t according to whether the original organism did or did not split in the small interval $(0, \delta t)$. If it did not, then $p_{1n}(t - \delta t)$ is the probability of then growing to n in the residual time $t - \delta t$ while if it did, then $p_{2n}(t - \delta t)$ is the probability of growing to n in the remaining time $t - \delta t$. That is, for $n \geqslant 2$,

$$p_{1n}(t) = [\lambda \delta t + o(\delta t)] \, p_{2n}(t - \delta t) + [1 - \lambda \delta t + o(\delta t)] \, p_{1n}(t - \delta t).$$

This means that

$$\frac{p_{1n}(t) - p_{1n}(t - \delta t)}{\delta t} = \lambda p_{2n}(t - \delta t) - \lambda p_{1n}(t - \delta t) + \frac{o(\delta t)}{\delta t}$$

and, in the limit, as $\delta t \to 0$,

$$\frac{dp_{1n}(t)}{dt} = \lambda p_{2n}(t) - \lambda p_{1n}(t), n \geqslant 2. \tag{2.3}$$

When $n = 1$, there has been no change in the interval $(0, t)$ and $p_{11}(t) = [1 - \lambda \delta t + o(\delta t)] \, p_{11}(t - \delta t)$, which leads to

$$\frac{dp_{11}(t)}{dt} = -\lambda p_{11}(t). \tag{2.4}$$

It is seen however that equation (2.4) is but a particular instance of (2.3) provided we set $p_{21}(t) \equiv 0$, which is reasonable enough since the number cannot decrease.

Perhaps the easiest way to solve the equation (2.3) is to use a generating function. Suppose that,

$$G_1(\theta, t) = \sum_{n=1}^{\infty} p_{1n}(t) \theta^n, |\theta| \leqslant 1.$$

Multiply (2.3) by θ^n and sum from 1 to ∞. This gives

$$\sum_{n=1}^{\infty} \frac{dp_{1n}(t)}{dt} \theta^n$$

$$= \lambda \sum_{1}^{\infty} p_{2n}(t)\theta^n - \lambda \sum_{1}^{\infty} p_{1n}(t)\theta^n,$$

$$= \lambda \sum_{1}^{\infty} p_{2n}(t)\theta^n - \lambda G_1(t,\theta). \tag{2.5}$$

Now $p_{2n}(t)$ has an interesting relationship to $p_{1n}(t)$. Evidently those n present at time t are descendants of one or the other the two originally present. Now the probability generating function of the sum of independent random variables is the product of their probability generating functions. (See Appendix 1.) But $\sum_{1}^{\infty} p_{2n}(t)\theta^n$ is the probability generating function, $G_2(\theta,t)$, for the number present at time t when the colony has two initiators. Hence this must be the product of the probability generating functions for the numbers of descendants from each of the originators. That is to say, $G_2(\theta,t) = G_1(\theta,t)G_1(\theta,t) = G_1^2(\theta,t)$. Thus (2.5) can be cast in the form of a simple partial differential equation. That is to say, writing $G_1(\theta,t) = G_1$.

$$\frac{\partial G_1}{\partial t} = \lambda G_1^2 - \lambda G_1,$$

$$\left[\frac{1}{G_1(G_1-1)}\right]\frac{\partial G_1}{\partial t} = \lambda,$$

$$\left(\frac{1}{G_1-1} - \frac{1}{G_1}\right)\frac{\partial G_1}{\partial t} = \lambda,$$

$$\frac{\partial}{\partial t}[\log(1-G_1) - \log G_1] = \lambda,$$

$$\frac{\partial}{\partial t}\{\log[(1-G_1)/G_1]\} = \lambda. \tag{2.6}$$

Integrating with respect to time,

$$\log[(1-G_1)/G_1] = \lambda t + c(\theta),$$

where the 'constant' $c(.)$ may involve θ, since (2.6) involves a partial derivative with respect to time. Thus

$$(1-G_1)/G_1 = e^{\lambda t}k(\theta). \tag{2.7}$$

$k(\theta) = \exp c(\theta)$ must be found from the initial conditions. But we know that originally there was one organism — that is $p_{1n}(0) = 0(n \neq 1)$ and $p_{11}(0) = 1$. In terms of $G_1(\theta,t)$ this means that $G_1(\theta,0) = \theta p_{11}(0) = \theta$. Setting $t = 0$ in (2.7) we have $(1-\theta)/\theta = k(\theta)$. We may now solve for G_1, arriving at

$$G_1(\theta,t) = \theta/[\theta+(1-\theta)\exp(\lambda t)]. \qquad (2.8)$$

We are now able to recover $p_{1n}(t)$ by expanding $G_1(\theta,t)$ as a power series in θ and identifying the coefficient of θ^n. We write G_1 in the form,

$$G_1(\theta,t) = \frac{\theta \exp(-\lambda t)}{1-\theta[1-\exp(-\lambda t)]}, \qquad (2.9)$$

and since $1-\exp(-\lambda t) < 1$, and $|\theta| < 1$

$$G_1(\theta,t) = \theta \exp(-\lambda t)\{1-\theta[1-\exp(-\lambda t)]\}^{-1}$$

$$= \theta \exp(-\lambda t)\sum_{r=0}^{\infty}[\theta\{1-\exp(-\lambda t)\}]^r.$$

The coefficient of θ^n is clearly

$$p_{1n}(t) = [1-\exp(-\lambda t)]^{n-1}\exp(-\lambda t), \; n \geqslant 1. \qquad (2.10)$$

Problem 1. Solve the equation (2.4) directly. Assuming (2.10) for $n = 1,2,\ldots,r-1$, show that $p_{2r}(t) = (r-1)[1-\exp(-\lambda t)]^{r-2}\exp(-2\lambda t)$. Use (2.3) to solve for $p_{1r}(t)$. ∎

Problem 2. Show that $G_1[G_1(\theta,t),t] = G_1(\theta,2t)$. By considering adjoining intervals of duration t, find a probabilistic proof of this result. ∎

Much additional information can be squeezed from $G_1(\theta,t)$. Suppose we require the expected number of organisms present at time t when initially one was present. This is

$$\sum_{n=1}^{\infty}np_{1n}(t) \text{ but is also } \left[\frac{\partial G_1(\theta,t)}{\partial\theta}\right]_{\theta=1} = e^{\lambda t},$$

easily from (2.9). Thus the expected number tends to infinity as t increases. As it happens, for this situation, we can obtain the expected number of organisms by noting that the number present has a geometric distribution with parameter $\exp(-\lambda t)$, whence the expected number is $\exp(\lambda t)$.

Problem 3. Show that the variance of the number of organisms at time t is $\exp(2\lambda t)[1-\exp(-\lambda t)]$ by operating on the probability generating function $G_1(\theta,t)$. ∎

We may, without additional machinery, deal with the case when initially there were m organisms. For each of those present at time t is the descendant of just one original ancestor. Hence the number present is the sum of those descendants and the appropriate p.g.f. is the mth power of $G_1(\theta,t)$. That is to say

$$G_m(\theta,t) = [G_1(\theta,t)]^m. \tag{2.11}$$

Problem 4. Show, by using equation (2.11), that

$$p_{mn}(t) = \binom{n-1}{n-m} \exp(-m\lambda t) \, [1-\exp(-\lambda t)]^{n-m}, \, n \geqslant m. \quad \blacksquare$$

Problem 5. Use a probabilistic argument to show that, for arbitrary fixed u such that $0 < u < t$,

$$p_{mn}(t) = \sum_{r=m}^{n} p_{mr}(u)p_{rn}(t-u). \quad \blacksquare$$

Problem 6. From equation (2.3) deduce that

$$p_{1n}(t) = \exp(-\lambda t) \int_0^t \lambda \exp(\lambda u) p_{2n} (u) \, du, n \geqslant 2.$$ Obtain an alternative probabilistic argument. $\blacksquare$

2.3 FORWARD AND BACKWARD EQUATIONS

The pure birth process can also be explored from the point of view of what happens immediately after time t. There will be n organisms present at time $t + \delta t$ either if there were n at time t and no birth took place in the next interval $(t,t+\delta t)$, or if there were $n-1$ present at time t and just one birth takes place in the subsequent interval $(t,t+\delta t)$. For $n \geqslant 2$

$$p_{1n}(t+\delta t) = p_{1n}(t) \, [1-\lambda\delta t+o(\delta t)] \,^n + p_{1,n-1}(t) \binom{n-1}{1} [\lambda\delta t+o(\delta t)]$$

$$[1-\lambda\delta t+o(\delta t)]^{n-2}$$

$$= p_{1n}(t) \, [1-n\lambda\delta t+o(\delta t)] + p_{1,n-1}(t) \, (n-1) \, [\lambda\delta t+o(\delta t)].$$

We observe that $[1-\lambda\delta t + o(\delta t)]^n$ is the probability that not one of the n organisms yields a birth in duration δt and $\binom{n-1}{1} [\lambda\delta t + o(\delta t)] \, [1-\lambda\delta t + o(\delta t)]^{n-2}$ is the probability that just one of $n-1$ provides a birth. Other possibilities are conceivable but have negligible probability. We now have

$$\frac{p_{1n}(t+\delta t)-p_{1n}(t)}{\delta t} = -n\lambda p_{1n}(t) + (n-1)\lambda p_{1,n-1}(t) + \frac{o(\delta t)}{\delta t}, n \geqslant 2,$$

and, proceeding to the limit as $\delta t \to 0$,

$$\frac{dp_{1n}(t)}{dt} = -n\lambda p_{1n}(t) + (n-1)\lambda p_{1,n-1}(t), \ n \geq 2. \tag{2.12}$$

The corresponding equation for $n = 1$ is

$$\frac{dp_{11}(t)}{dt} = -\lambda p_{11}(t), \tag{2.13}$$

which agrees with (2.12) with $p_{10}(t) \equiv 0$.

Technically these are known as the *forward* equations, since they look 'forward' from t to $t + \delta t$, as opposed to the previous set, which are called the *backward* equations. In the 'backward' equations, the final number of organisms present is constant while in the 'forward' equations, the initial number is fixed.

Problem 7. Verify that $p_{1n}(t) = [1-\exp(-\lambda t)]^{n-1} \exp(-\lambda t), (n \geq 1)$ satisfies the forward equations, (2.12). ∎

Problem 8. (Optional.) One approach to the solution of differential equations is by the method of Laplace transforms. Recall that for functions of non-negative variables, t, the Laplace transform of $\phi(t)$ is defined as

$$\phi^*(s) = \int_0^\infty \phi(t)e^{-st}dt. \quad \text{(See Appendix 2.)}$$

Show that the Laplace transform of the forward equations (2.12) is

$$-p_{1n}(0) + sp^*_{1n}(s) = \lambda(n-1)p^*_{1,n-1}(s) - \lambda np^*_{1n}(s), n \geq 1.$$

If $p_{1n}(t) = [1-\exp(-\lambda t)]^{n-1}\exp(-\lambda t)$, show that $p^*_{1n}(s)$ satisfies this last equation. ∎

Problem 9. Assuming that the equation (2.12) has a solution of the form

$$p_{1n}(t) = \sum_{i=1}^{n} A_n^{(i)} \exp(-i\lambda t), \text{ show that}$$

$$A_n^{(i)} = \binom{n-1}{i-1} A_i^{(i)}.$$

Using the initial condition, $p_{1n}(0) = 0, n \neq 1, p_{11}(0) = 1$, show that $A_i^{(i)} = (-1)^{i+1}$ and hence that $p_{1n}(t) = (1-e^{-\lambda t})^{n-1} e^{-\lambda t}$. ∎

Our standard technique for solving an infinite set of differential equations such as (2.12) is to throw them together into a partial differential equation involving the probability generating function $G_1(\theta,t) = \sum_{n=1}^{\infty} p_{1n}(t)\,\theta^n$. Multiplying (2.12) by θ^n and summing over n we have

$$\sum_{1}^{\infty} \frac{dp_{1n}(t)\theta^n}{dt} = -\lambda \sum_{1}^{\infty} np_{1n}(t)\theta^n + \lambda \sum_{1}^{\infty} (n-1)p_{1,n-1}(t)\theta^n, \text{ or}$$

$$\frac{\partial}{\partial t} \sum_{1}^{\infty} p_{1n}(t)\theta^n = -\lambda\theta \sum_{1}^{\infty} p_{1n}(t)\frac{d}{d\theta}(\theta^n) + \lambda\theta^2 \sum_{2}^{\infty} p_{1,n-1}(t)\frac{d}{d\theta}(\theta^{n-1}).$$

That is to say

$$\frac{\partial}{\partial t}[G_1(\theta,t)] = -\lambda\theta\frac{\partial}{\partial\theta}[G_1(\theta,t)] + \lambda\theta^2\frac{\partial}{\partial\theta}[G_1(\theta,t)]. \qquad (2.14)$$

This is not a course in differential equations though the solution by the method of Lagrange is followed up in (optional) Problem 10. We content ourselves with the remark that

$$G_1(\theta,t) = K[\theta \exp(-\lambda t)/(1-\theta)]$$

is always a solution for any function $K(.)$ and that the initial condition $G_1(\theta,0) = \theta$ forces

$$G_1(\theta,t) = \theta\exp(-\lambda t)/\{1-\theta[1-\exp(-\lambda t)]\}.$$

Problem 10. (Optional.) For the partial differential equation, the auxiliary equations are

$$\frac{dt}{1} = \frac{d\theta}{\lambda\theta(1-\theta)} = \frac{dG_1}{0}.$$

Show that $G_1 = a$, $\theta\exp(-\lambda t)/(1-\theta) = b$ are solutions and hence that the general solution is

$$G_1(\theta,t) = K[\theta\exp(-\lambda t)/(1-\theta)],$$

where $K(.)$ is an arbitrary function. Solve for $K(.)$ when $G_1(\theta,0) = \theta$. ∎

Time to the nth Birth

The number of births in time t never decreases hence if T_n is the time to the nth birth and there was one originator

$$Pr[T_n \leqslant t] = Pr \text{ [at least } n+1 \text{ present at time } t]$$

$$= \sum_{k=n+1}^{\infty} [1-\exp(-\lambda t)]^{k-1} \exp(-\lambda t), \text{ using (2.10),}$$

$$= [1-\exp(-\lambda t)]^n.$$

But this is the *cumulative distribution* function of T_n, hence the probability density function is

$$\frac{d}{dt}[Pr(T_n \leqslant t)] = n \, [1-\exp(-\lambda t)]^{n-1} \, \lambda \exp(-\lambda t). \tag{2.15}$$

In considering the time to the nth birth, as noted in section 2.1, when there are k organisms present, the waiting times for each of these to provide a birth have independent exponential distributions, each with parameter λ. The time to the *next* birth is thus the minimum of these and has *another* exponential distribution with parameter $k\lambda$. This structural result may be exploited in various ways. For example

$$T_n = (T_n - T_{n-1}) + (T_{n-1} - T_{n-2}) \ldots + (T_2 - T_1) + T_1,$$

$$E(T_n) = \frac{1}{n\lambda} + \frac{1}{(n-1)\lambda} + \ldots + \frac{1}{\lambda}.$$

Problem 11. Derive (2.15) directly by noting that T_n will fall in the interval $(t, t + \delta t)$ if there are n present at time t and the next birth falls in $(t, t + \delta t)$. Use (2.15) to obtain $E(T_n)$. ∎

Problem 12. Suppose initially there are two organisms, one of type A, the other of type B. If these two independently generate colonies in pure birth processes with different birth rates λ_A, λ_B, calculate the expected size of the type A colony when the first birth of type B occurs.

2.4 BIRTH AND DEATH PROCESS

Suppose, independently for each organism, at time t the probability of giving birth to just one more organism in the interval $(t, t+\delta t)$ is $\lambda \delta t + o(\delta t)$ and the probability of dying is $\mu \delta t + o(\delta t)$. These are taken to be mutually exclusive events and we further assume that the probability that nothing happens is $1 - (\lambda + \mu) \, \delta t + o(\delta t)$. We find a backward equation for $p_{1n}(t)$, the probability that there are $n \geqslant 1$ organisms present at time t, when initially one was present. There is either a birth in the interval $(0, \delta t)$ or nothing happens in the interval $(0, \delta t)$, and there are eventually n present after a further duration of $t - \delta t$. Hence

$$p_{1n}(t) = [\lambda \delta t + o(\delta t)] \, p_{2n}(t-\delta t) + [1 - \lambda \delta t - \mu \delta t + o(\delta t)] \, p_{1n}(t-\delta t) + o(\delta t),$$

which leads to

$$\frac{dp_{1n}(t)}{dt} = -(\lambda + \mu)p_{1n}(t) + \lambda p_{2n}(t), \quad n \geqslant 1. \tag{2.16}$$

The case $n = 0$ is special, since if the initial organism dies in the interval $(0, \delta t)$, then it is certain that there is no organism present at any later time!

$$p_{10}(t) = [\lambda \delta t + o(\delta t)] \, p_{20}(t - \delta t) + [1 - \lambda \delta t - \mu \delta t + o(\delta t)] \, p_{10}(t - \delta t)$$
$$+ \mu \delta t + o(\delta t).$$

$$\frac{dp_{10}(t)}{dt} = -(\lambda + \mu)p_{10}(t) + \lambda p_{20}(t) + \mu. \tag{2.17}$$

If $G_1(\theta, t) = \sum_{n=0}^{\infty} p_{1n}(t) \, \theta^n$, then multiplying (2.16) by θ^n, summing from $n = 1$ to ∞ and adding on (2.17) we arrive at

$$\frac{\partial G_1}{\partial t} = -(\lambda + \mu)G_1 + \lambda \sum_{n=0}^{\infty} p_{2n}(t)\theta^n + \mu, \tag{2.18}$$

where $G_1 = G_1(\theta, t)$.

But the surviving number present if initially there were two is the sum of the survivors of each of these originators. That is

$$G_2(\theta, t) = \sum_0^{\infty} p_{2n}(t) \theta^n = [G_1(\theta, t)]^2.$$

Hence,

$$\frac{\partial G_1}{\partial t} = -(\lambda + \mu)G_1 + \lambda G_1^2 + \mu,$$

or

$$\frac{\partial G_1}{\partial t} = (\lambda G_1 - \mu)(G_1 - 1), \tag{2.19}$$

which of course reduces to the pure birth process if $\mu = 0$.
We can re-write this equation as

$$\left[\frac{1}{(\lambda G_1 - \mu)(G_1 - 1)} \right] \frac{\partial G_1}{\partial t} = 1,$$

$$\frac{1}{(\mu - \lambda)} \left[\frac{\lambda}{\lambda G_1 - \mu} - \frac{1}{G_1 - 1} \right] \frac{\partial G_1}{\partial t} = 1.$$

Hence, when $\mu > \lambda$, we have

$$\frac{\partial}{\partial t} \left[\log \left\{ (\lambda G_1 - \mu) / (G_1 - 1) \right\} \right] = \mu - \lambda,$$

which gives

$$\log \left[(\lambda G_1 - \mu) / (G_1 - 1) \right] = (\mu - \lambda)t + c(\theta).$$

Thus

$$(\lambda G_1 - \mu) \, / \, (G_1 - 1) = \exp \, [(\mu - \lambda)t] \, k \, (\theta), \qquad (2.20)$$

where $k(\theta)$ depends only on θ.

Now when $t = 0$, $p_{1n}(0) = 0$, $n \neq 1$, $p_{11}(0) = 1$. These initial conditions imply $G_1(\theta, 0) = \theta$, which in turn implies $k(\theta) = (\lambda \theta - \mu) \, / \, (\theta - 1)$. Substituting for $k(\theta)$ in (2.20).

$$\frac{\lambda G_1 - \mu}{G_1 - 1} = \left(\frac{\lambda \theta - \mu}{\theta - 1} \right) \exp \, [(\mu - \lambda)t].$$

This is easily solved for G_1 giving

$$G_1 \, (\theta, t) = \frac{\mu(1 - \theta) - (\mu - \lambda \theta) \exp \, [(\mu - \lambda)t]}{\lambda(1 - \theta) - (\mu - \lambda \theta) \exp \, [(\mu - \lambda)t]}. \qquad (2.21)$$

At the end of such a computation, it is important to check that boundary conditions such as $G_1(1, t) \equiv 1$, $G_1(\theta, 0) \equiv \theta$ are satisfied. Equation (2.21) similarly holds for $\mu < \lambda$. For $\mu = \lambda$, see Problem 16.

Some new questions arise in this process. We may ask, 'What is the probability that at time t, there are no organisms present?' This must be $p_{10}(t) = G_1(0, t)$ which, if $\lambda \neq \mu$, is

$$\frac{\mu - \mu \exp \, [(\mu - \lambda)t]}{\lambda - \mu \exp \, [(\mu - \lambda)t]}. \qquad (2.22)$$

We remark that (2.22) is the probability that the colony is extinct *at or before* t. If $\mu < \lambda$, $p_{10}(t) \to \mu/\lambda$ as $t \to \infty$, while if $\mu > \lambda$, $p_{10}(t) \to 1$ as $t \to \infty$, i.e. eventual extinction is certain.

Problem 13. For $\lambda = \mu$, show that the probability of extinction at or before t, is $\lambda t/(1 + \lambda t)$. (Hint: expand the numerator and denominator of (2.22) as power series and let $\lambda \to \mu$ or employ the ever-resourceful rule of L'Hôpital.) ■

Problem 14. If $\mu > \lambda$, extinction is certain and (2.22) must be the distribution function of the time, T_0, to extinction. Show that

$$E(T_0) = (1/\lambda) \log \, [\mu/(\mu - \lambda)]. \quad ■$$

Problem 15. Show that the expected number of organisms present at time t is $\exp \, [(\lambda - \mu)t]$, provided $\lambda \neq \mu$. ■

Problem 16. If $\lambda = \mu$, show that

$$G_1(\theta,t) = [\lambda t + \theta (1 - \lambda t)] / (1 + \lambda t - \theta \lambda t)$$

and hence

$$p_{1n}(t) = (\lambda t)^{n-1} / (1 + \lambda t)^{n+1}, \quad n \geqslant 1,$$

$$p_{10}(t) = \lambda t / (1 + \lambda t). \quad \text{(See also Problem 13.)} \quad \blacksquare$$

Problem 17. If originally there are m such organisms, calculate the expected number present at time t, and the probability of extinction by time t. (Hint: Consider the descendants of each of the m originators.) $\quad \blacksquare$

The particular case $\lambda = 0, \mu > 0$ is interesting and is one of **pure death**.

Problem 18. Show that if there are m organisms originally, then in the pure death process, for which $\lambda = 0, \mu > 0$,

$$G_m(\theta,t) = [1 + (\theta - 1) \exp(-\mu t)]^m$$

and hence that

$$p_{mn}(t) = \binom{m}{n} e^{-n\mu t} (1 - e^{-\mu t})^{m-n}, \quad n \leqslant m. \quad \blacksquare$$

Problem 19. (Harder.) If initially there are m males and $k-m$ females both subject to pure death process with the same $\mu(> 0)$, find the expected number of surviving males when the females become extinct. $\quad \blacksquare$

2.5 DEATH AND IMMIGRATION PROCESS

We may pursue a slightly different model by suppressing births in the birth and death process but allowing random arrivals from outside the colony. More formally, every organism present in the colony has a probability $\mu \delta t + o(\delta t)$ of dying in the interval $(t, t + \delta t)$ regardless of the time and of the history of any other organism. Moreover, there is a probability $\nu \delta t + o(\delta t)$ that a single organism joins the colony as an immigrant and $1 - \nu \delta t + o(\delta t)$ of no immigrant in the interval $(t, t + \delta t)$. It happens that it is easier to solve the forward equations. If initially there are m organisms then there will be n at time $t + \delta t$ if there were $n-1, n$ or $n+1$ at time t and one addition, no change or one death in the interval $(t, t + \delta t)$. We recover the following equation $(n \geqslant 1)$

$$p_{mn}(t + \delta t) = [1 - \mu \delta t + o(\delta t)]^{n-1} [\nu \delta t + o(\delta t)] p_{m,n-1}(t)$$

$$+ [1 - \mu \delta t + o(\delta t)]^n [1 - \nu \delta t + o(\delta t)] p_{mn}(t)$$

$$+ \binom{n+1}{1} [\mu \delta t + o(\delta t)] [1 - \mu \delta t + o(\delta t)]^n [1 - \nu \delta t + o(\delta t)] p_{m,n+1}(t),$$

giving

$$p_{mn}(t+\delta t) = v\delta t\, p_{m,n-1}(t) + [1-(n\mu+v)\,\delta t]\, p_{mn}(t)$$
$$+ (n+1)\,\mu\delta t\, p_{m,n+1}(t) + o(\delta t).$$

Thus

$$\frac{dp_{mn}(t)}{dt} = -(v+n\mu)p_{mn}(t) + (n+1)\mu p_{m,n+1}(t) + vp_{m,n-1}(t). \tag{2.23}$$

It is easily verified that (2.23) holds also for $n = 0$, so long as $p_{m,-1}(t) \equiv 0$.

Let $G(\theta,t) = \sum\limits_{n=0}^{\infty} p_{mn}(t)\,\theta^n$, multiply the differential equation (2.23) by θ^n and sum over n. We have

$$\sum_{n=0}^{\infty} \theta^n \frac{dp_{mn}(t)}{dt} = \frac{\partial}{\partial t}\left[\sum_{n=0}^{\infty} \theta^n p_{mn}(t)\right] = \frac{\partial}{\partial t}\,[G(\theta,t)].$$

$$\sum_{0}^{\infty} (v+n\mu)\theta^n p_{mn}(t) = vG(\theta,t) + \mu\theta \sum_{1}^{\infty} (\frac{d}{d\theta}\,\theta^n)p_{mn}(t)$$

$$= vG(\theta,t) + \mu\theta \frac{\partial}{\partial\theta}\,[G(\theta,t) - p_{m0}(t)]$$

$$= vG(\theta,t) + \mu\theta \frac{\partial}{\partial\theta}\,[G(\theta,t)].$$

$$\mu \sum_{0}^{\infty} (n+1)\theta^n p_{m,n+1}(t) = \mu \sum_{0}^{\infty} \frac{d}{d\theta}\,\theta^{n+1} p_{m,n+1}(t) = \mu \frac{\partial}{\partial\theta}\,[G(\theta,t)].$$

$v \sum\limits_{0}^{\infty} \theta^n p_{m,n-1}(t) = v\theta G(\theta,t)$. Collecting together all the terms,

$$\frac{\partial G}{\partial t} = -(vG+\mu\theta \frac{\partial G}{\partial\theta}) + \mu \frac{\partial G}{\partial\theta} + v\theta G,$$

$$= v(\theta-1)G - \mu(\theta-1)\frac{\partial G}{\partial\theta}.$$

The solution to this equation, after incorporating the boundary condition $p_{mn}(0) = 0\,(n \neq m), p_{mn}(0) = 1$ is

$$G(\theta,t) = \exp\left[\frac{v}{\mu}(\theta-1)(1-e^{-\mu t})\right]\left[1+(\theta-1)e^{-\mu t}\right]^m. \tag{2.24}$$

We note several interesting features of this solution. When $m = 0$, the probability generating function corresponds to a Poisson distribution, with parameter $\frac{v}{\mu}[1-\exp(-\mu t)]$. Evidently when $m \neq 0$, the number present is the sum of the survivors of the original m — who are subject to a pure death process — and those immigrants still present. Since these two groups behave independently of each other, the full probability generating function is the product of the corresponding separate probability generating functions. At once we have that the expected number present at time t is the mean of the Poisson distribution for $m = 0$ plus the mean of the pure death process, that is,

$$\frac{\nu}{\mu} [1 - \exp(-\mu t)] + m \exp(-\mu t).$$

Problem 20. State the probability that just r immigrant organisms are in the colony at time t. State the probability that just r of the original m organisms remain in the region at time t. Calculate also the expected number of deaths by time t. (Note that the number of arrivals in duration t *also* has a Poisson distribution.) ∎

Problem 21. (Optional.) To solve the partial differential equation (method of Lagrange)

$$\frac{\partial G}{\partial t} = \nu(\theta - 1)G - \mu(\theta - 1)\frac{\partial G}{\partial \theta}.$$

From the associated equations

$$\frac{dt}{1} = \frac{d\theta}{\mu(\theta - 1)} = \frac{dG}{\nu(\theta - 1)G},$$

show that the form of the solution must be

$G(\theta,t) = e^{\frac{\nu\theta}{\mu}} K[(\theta - 1)e^{-\mu t}]$, where $K(.)$ is a general function. By using the boundary conditions $G(\theta,0) = \theta^m$, show that

$$K(y) = (1+y)^m \exp\left[-\frac{\nu}{\mu}(y+1)\right],$$

and hence recover the probability generating function for the death–immigration process. ∎

We may also study the limiting behaviour for the death–immigration process for large t. Regardless of the initial number present,

$$\lim_{t \to \infty} G(\theta,t) = \exp\left[\frac{\nu}{\mu}(\theta - 1)\right].$$

That is, for large t, the number of organisms present has a Poisson distribution with parameter ν/μ. If we are only interested in the form of the limiting distribution, *supposing it to exist*, then we need not solve for $G(\theta,t)$ at all. For suppose $p_{mn}(t) \to p_{mn}$ then $\dfrac{dp_{mn}(t)}{dt} \to 0$ and $\dfrac{\partial G(\theta,t)}{\partial t} \to 0$ as $t \to \infty$. Hence, in the limit, we have

$$0 = \nu(\theta-1)G - \mu(\theta-1)\frac{dG}{d\theta};$$

that is,

$$\frac{1}{G} \cdot \frac{dG}{d\theta} = \frac{\nu}{\mu}, \text{ or } \frac{d}{d\theta}(\log G) = \frac{\nu}{\mu},$$

whence, easily, $G = \exp\left[\frac{\nu}{\mu}(\theta-1)\right]$. In fact, if we are prepared to assume the existence of a limit, then p_{mn} can be found without even forming the partial differential equation for $G(\theta,t)$. For the original differential equation was

$$\frac{dp_{mn}(t)}{dt} = -(\nu+n\mu)p_{mn}(t) + (n+1)\mu p_{m,n+1}(t) + \nu p_{m,n-1}(t).$$

If a limit exists, then

$$0 = -(\nu+n\mu)p_{mn} + (n+1)\mu p_{m,n+1} + \nu p_{m,n-1},$$

that is

$$\nu(p_{mn} - p_{m,n-1}) = \mu[(n+1)p_{m,n+1} - np_{mn})]. \qquad (2.25)$$

Sum both sides of (2.25) from $n=0$ to $n=r-1$; $\nu p_{m,r-1} = \mu r p_{mr}$ or $p_{mr} = (\nu/\mu r)p_{m,r-1}$. By repeated application,

$$p_{mr} = \left(\frac{\nu}{\mu}\right)^r \frac{1}{r!} p_{m0}.$$

But $\displaystyle\sum_{r=0}^{\infty} p_{mr} = 1$. Hence $1 = p_{m0}\exp(\nu/\mu)$ and $p_{mr} = \left(\dfrac{\nu}{\mu}\right)^r \dfrac{1}{r!}\exp(-\nu/\mu)$, which corresponds to a Poisson distribution with parameter ν/μ, whatever the value of m.

Problem 22. (Birth and immigration.) The probabilities that govern the behaviour of particles in a region of space satisfy the following conditions: particles behave independently of each other and for each particle the probability that it splits into two in $(t,t+\delta t)$ is $\lambda\delta t + o(\delta t)$ whatever the time. Independently of the state of the system, the probability that a new particle enters is $\nu\delta t + o(\delta t)$. Given that initially the region contains one particle, show that $p_n(t)$, the probability that there are n particles in the region at time t, satisfies the differential equation

$$\frac{dp_n(t)}{dt} = [\nu+(n-1)\lambda]p_{n-1}(t) - [\nu+n\lambda]p_n(t), \qquad n = 1,\ldots.$$

Hence, or otherwise, show that the probability generating function $G(\theta,t) = \sum_1^\infty \theta^n p_n(t)$ satisfies

$$\frac{\partial G}{\partial t} = \nu(\theta-1)G + \lambda\theta(\theta-1)\frac{\partial G}{\partial \theta} \,,$$

verify that, with $G(\theta,0) = \theta$,

$$G(\theta,t) = \frac{\theta \exp\left[-(\lambda+\nu)t\right]}{[1-\theta\{1-\exp(-\lambda t)\}]^{(\lambda+\nu)/\lambda}} \tag{2.26}$$

Hence or otherwise calculate (a) the expected number of particles present in the region at time t, (b) the probability that there are just n particles in the region at time t in the case $\lambda = \nu$.

REFERENCES

[1] *Mathematical Models in the Social, Management and Life Sciences*, D. N. Burghes and A. D. Wood. Ellis Horwood Ltd., Chichester, 1980.
[2] G. U. Yule, *Ph. Tr. Royal Society, London, Series B*, **213** (1924).
[3] W. H. Furry, *Physical Reviews*, **52** (1937).

BRIEF SOLUTIONS AND COMMENTS ON THE PROBLEMS

Problem 1. From $\frac{\mathrm{d}p_{11}(t)}{\mathrm{d}t} = -\lambda p_{11}(t)$, which can be written $\frac{\mathrm{d}}{\mathrm{d}t}\left[\log\{p_{11}(t)\}\right]$ $= -\lambda$. After integrating, and using the initial condition $p_{11}(0) = 1$, we have $p_{11}(t) = \exp(-\lambda t)$. $p_{2r}(t) = \sum_{n=1}^{r-1} p_{1n}(t)p_{1,r-n}(t) = (r-1)\left[1-\exp(-\lambda t)\right]^{r-2}$ $\exp(-2\lambda t)$, after substituting for $p_{1n}(t), p_{1,r-n}(t)$. We may write (2.3) as

$$\frac{\mathrm{d}}{\mathrm{d}t}\left[\exp(\lambda t)p_{1r}(t)\right] = \lambda(r-1)\left[1-\exp(-\lambda t)\right]^{r-2}\exp(-\lambda t),$$

which may be integrated at sight to give $p_{1r}(t)$. The reader will detect the basis of a proof by induction. He may care to confirm $p_{2r}(t)$ by picking out the appropriate term in the expansion of $[G_1(\theta,t)]^2$.

Problem 2. It is slightly more convenient to write $G_1(\theta,t)$ as $\theta/[\theta+(1-\theta)$ $\exp(\lambda t)]$. Now substitute $G_1(\theta,t)$ for θ to obtain $\theta/[\theta+(1-\theta)\exp(2\lambda t)] = G(\theta,2t)$. Alternatively, at the end of time t there are m organisms present with probability $p_{1m}(t)$. The p.g.f. of the number present after a further interval t is then $G_m(\theta,t) = [G_1(\theta,t)]^m$. Hence the unconditional p.g.f. of the number after

2t is $\sum\limits_{m=1}^{\infty} [G(\theta,t)]^m p_{1m}(t) = G_1 [G_1(\theta,t),t]$. See Appendix 1.

Problem 3. To avoid excessive manipulation it may pay to cast $G_1(\theta,t)$ in the form

$$\theta \{[\exp(\lambda t) + \theta (1-\exp(\lambda t)]\}^{-1}.$$

After differentiating twice, with respect to θ, and setting $\theta = 1$, we obtain $2 \exp(2\lambda t) - 2 \exp(\lambda t)$, which is $E[N_t (N_t-1)]$. Now use $E(N_t)$, where N_t is the number of organisms present at time t.

Problem 4

$$G_m(\theta,t) = \theta^m \exp(-\lambda mt) \{1-\theta [1-\exp(-\lambda t)]\}^{-m}.$$

Expand the last factor as a power series and pick out the coefficient of θ^{n-m}.

Problem 5. At time u, there will be $r(\geqslant m)$, with probability $p_{mr}(u)$. For the remaining time $t-u$, the process behaves as if it started with r and then attains n with probability $p_{rn}(t-u)$. The alternative of summing the appropriate terms from Problem 4 does not look particularly inviting.

Problem 6. Multiply both sides of the equation by $\exp(\lambda t)$ and obtain

$$\frac{d}{dt} [\exp(\lambda t)p_{1n}(t)] = \lambda \exp(\lambda t)p_{2n}(t).$$

For $n \geqslant 2$, there is a first birth at $t-u$ and then the two become n after a further time u. Remember to integrate out u.

Problem 7. Straight forward differentiation and identification gives the result required; the task is eased by noting that $\exp(-2\lambda t) = \exp(-\lambda t) - \exp(-\lambda t) [1-\exp(-\lambda t)]$.

Problem 8. Integrating by parts,

$$\int_0^{\infty} \frac{dp_{11}(t)}{dt} \cdot \exp(-st)dt = \left[p_{1n}(t) \exp(-st) \right]_0^{\infty} + s \int_0^{\infty} p_{1n}(t) \exp(-st)dt$$

$$= -p_{1n}(0) + sp_{1n}^*(s).$$

The other side of the equation is immediate.
Write down the definition of $p_{1n}^*(s)$. After integrating by parts, for $n \geqslant 2$,

$$p_{1n}^*(s) = s \int_0^{\infty} \frac{[1-\exp(-\lambda t)]^n}{n\lambda} \exp(-st)dt.$$

Now write $[1-\exp(-\lambda t)]^n = [1-\exp(-\lambda t)]^{n-1} - [1-\exp(-\lambda t)]^{n-1}\exp(-\lambda t)$
Remember $p_{1n}(0) = 0$ for $n \geqslant 2$.

Problem 9. After substituting the suggested form for $p_{1n}(t)$ in (2.12), and equating coefficients of $\exp(-i\lambda t)$, we have

$$(n-i)A_n^{(i)} \equiv (n-1)A_{n-1}^{(i)}.$$

Thence the required result by repeated application. Since $p_{11}(0) = 1, A_1^{(1)} = 1$, hence $A_i^{(i)}$ by induction. Hence

$$p_{1n}(t) = \sum_{i=1}^{n} \binom{n-1}{i-1}(-1)^{i+1} \exp(-i\lambda t) = [1-\exp(-\lambda t)]^{n-1} \exp(-\lambda t).$$

Although the intention of Problem 7 is to verify that the known solution of the backward equations also satisfies the forward equations, we can of course use the method of induction directly. For this we write (2.12) as

$$\frac{d}{dt}[\exp(n\lambda t)p_{1n}(t)] = \frac{d}{dt}[\{\exp(\lambda t)-1\}^{n-1}].$$

Problem 10. We have immediately $G_1 = a$. From

$$\lambda dt = d\theta\left(\frac{1}{\theta} + \frac{1}{1-\theta}\right)$$

on integrating we obtain

$$\lambda t + c = \log[\theta/(1-\theta)]$$

or

$$be^{\lambda t} = \theta/(1-\theta).$$

Hence $G_1(\theta,t) = K[\theta \exp(-\lambda t)/(1-\theta)]$. To fit the initial condition $G_1(\theta,0) = \theta$, put $t = 0$, yielding $K[\theta/(1-\theta)] = \theta$. That is to say, $K(x) = x/(1+x)$, and, finally, set $x = \theta \exp(-\lambda t)/(1-\theta)$.

Problem 11. From equation (2.10), the probability of n present by time t is $[1-\exp(-\lambda t)]^{n-1}\exp(-\lambda t)$. Regardless of the arrival time of the last organism to appear, the time to the next arrival is exponential with parameter $n\lambda$. Hence the probability of one arrival in $(t,t+\delta t)$ is $n\lambda\delta t+o(\delta t)$. Therefore,

$$E(T_n) = \int_0^\infty t\ [1-\exp(-\lambda t)]^{n-1}\exp(-\lambda t)n\lambda dt = \int_0^\infty [1-\{1-\exp(-\lambda t)\}^n]\,dt.$$

Note,

$$1-[1-\exp(-\lambda t)]^n = \exp(-\lambda t) \sum_{r=0}^{n-1} [1-\exp(-\lambda t)]^r.$$

Problem 12. Suppose the first birth of type B is at time t. The expected number in the colony of type A is then $\exp(\lambda_A t)$. But the time to the first birth of type B has an exponential distribution say with parameter λ_B. Hence required expectation is

$$\int_0^\infty \lambda_B e^{(\lambda_A - \lambda_B)t} \, dt.$$

If $\lambda_B > \lambda_A$, this integral has value $\lambda_B/(\lambda_B - \lambda_A)$. If $\lambda_B \leqslant \lambda_A$, the integral diverges.

Problem 13. Expanding $\exp[\mu - \lambda)t]$ as $1 + (\mu - \lambda)t + \dots$, (2.22) becomes

$$\frac{\mu - \mu \, [1 + (\mu - \lambda)t \dots}{\lambda - \mu \, [1 + (\mu - \lambda)t \dots} \, ,$$

$$= \frac{\mu(\lambda - \mu)t + \dots \text{ higher powers in } (\lambda - \mu)}{(\lambda - \mu) + \mu(\lambda - \mu)t + \dots \text{ higher powers in } (\lambda - \mu)} \, ,$$

$$= \frac{\mu t + \dots}{1 + \mu t + \dots} \to \frac{\mu t}{1 + \mu t} \text{ as } \lambda \to \mu.$$

Or, differentiate numerator and denominator of (2.22) w.r.t. λ and let $\lambda \to \mu$.

Problem 14

$$E(T_0) = \int_0^\infty [1 - p_{10}(t)] \, dt$$

$$= \int_0^\infty \frac{\lambda - \mu}{\lambda - \mu \exp[(\mu - \lambda)t]} \, dt = \int_0^\infty \frac{(\mu - \lambda) \exp[-(\mu - \lambda)t]}{\mu - \lambda \exp[-(\mu - \lambda)t]} \, dt$$

$$= \frac{1}{\lambda} \left[\log\{\mu - \lambda e^{-(\mu - \lambda)t}\} \right]_0^\infty$$

$$= \frac{1}{\lambda} \log\left(\frac{\mu}{\mu - \lambda}\right)$$

Problem 15. We require $\left[\dfrac{\partial G_1(\theta, t)}{\partial \theta}\right]_{\theta = 1}$. Equation (2.21) does not appear specially attractive. However,

$$G_1(\theta, t) - 1 = \frac{(\mu - \lambda)(1 - \theta)}{\lambda(1 - \theta) - (\mu - \lambda\theta) \exp[(\mu - \lambda)t]} \, .$$

In differentiating this quotient, the factor $1-\theta$ simplifies the value at $\theta = 1$.

Problem 16. Using (2.21), to avoid 0/0, expand $\exp[(\mu-\lambda)t]$.

$$G_1(\theta,t) = \frac{\mu(1-\theta)-(\mu-\lambda\theta)\;[1+(\mu-\lambda)t\dots]}{\lambda(1-\theta)-(\mu-\lambda\theta)\;[1+(\mu-\lambda)t\dots]}.$$

After collecting terms, dividing out $\lambda-\mu$, and then $\mu \to \lambda$, the result is obtained. Alternatively return to equation (2.19) with $\mu = \lambda$.

$$\lambda = \frac{1}{(G_1-1)^2}\frac{\partial G_1}{\partial t} = -\frac{\partial}{\partial t}\left(\frac{1}{G_1-1}\right).$$

Then integrate and impose the boundary condition $G_1(\theta,0) = \theta$.

$$G_1(\theta,t) = 1 + \frac{\theta-1}{1+\lambda t-\theta\lambda t} = 1 + \frac{\theta-1}{1+\lambda t}\left(1-\frac{\theta\lambda t}{1+\lambda t}\right)^{-1}.$$

Now use the infinite binomial expansion to obtain the coefficient of θ^n.

Problem 17. From Problem 15, the expected number descended from each originator is $\exp[(\lambda-\mu)t]$, $\lambda \neq \mu$. Hence $m\exp[(\lambda-\mu)t]$ is the number for m originators. The case $\lambda = \mu$ gives no trouble this time and is m, either on substituting $\lambda = \mu$ or calculating $\left(\dfrac{\partial G_1}{\partial \theta}\right)_{\theta=1}$ for the p.g.f. announced in Problem 16. If $\lambda \neq \mu$, the probability of extinction at or before t is $[p_{10}(t)]^m$, where $p_{10}(t)$ is given at (2.22). For $\lambda = \mu$, $p_{10}(t) = \lambda t/(1+\lambda t)$, see Problem 13.

Problem 18. Set $\lambda = 0$ in $G_1(\theta,t)$ as displayed in equation (2.21) to obtain $1+(\theta-1)\exp(-\mu t) = 1-\exp(-\mu t)+\theta\exp(-\mu t)$. Take the mth power for m originators and pick out coefficient of θ^n.

Problem 19. The probability that all $k - m$ females are extinct at or before t is $[1-\exp(-\mu t)]^{k-m} = p_0(t)$ say. The p.d.f. to the time of extinction is $\dfrac{d}{dt}p_0(t)$, when the expected number of males present is $m\exp(-\mu t)$. Hence the unconditional expected number of males is

$$m\int_0^\infty \exp(-\mu t)p_0'(t)dt = m/(k-m+1).$$

$y = 1-\exp(-\mu t)$ is a helpful substitution in the integral.

Problem 20. For the probability of just r immigrants, the number originally present is irrelevant and we put $m = 0$ in (2.24). Hence the probability of r has a

Poisson distribution with parameter $\tau = \nu\,[1-\exp(-\mu t)]/\mu$. Of the m originators, each has independent probability $\exp(-\mu t)$ of surviving for time t. Hence the second required probability is $\binom{m}{r}p^r(1-p)^{m-r}$, where $p = \exp(-\mu t)$. The expected number of arrivals by time t is νt. We have $m + \nu t -$ expected deaths = expected number present.

Problem 21. From $\mu dt = d\theta/(\theta-1)$, we obtain $a = (\theta-1)\exp(-\mu t)$. From $\nu d\theta = \mu dG/G$, $b = \exp(-\theta)G^{\mu/\nu}$. Hence, the general form of $G(\theta,t)$ is as stated. When $t = 0$, $\theta^m = \exp(\nu\theta/\mu)K(\theta-1)$. Now put $\theta = 1+y$.

Problem 22. There will be n present at time $t+\delta t$ if

(i)   there were n at t, none split and none entered, with probability $(1-\lambda\delta t)^n\,(1-\nu\delta t) = 1 - (n\lambda+\nu)\,\delta t+o(\delta t)$; or

(ii) there were $n-1$ at t, one splits but none entered, with probability $(n-1)\,\lambda\delta t\,(1-\lambda\delta t)^{n-2}\,(1-\nu\delta t) = (n-1)\,\lambda\delta t+o(\delta t)$; or

(iii) there were $n-1$ at t, none split and one entered, with probability $(1-\lambda\delta t)^{n-1}\,.\nu\delta t = \nu\delta t+o(\delta t)$.

The case $n = 1$ should be considered. Obtaining (2.26) is routine.

$$\frac{\partial G}{\partial \theta} = \frac{G}{\theta} + \frac{[\{1-\exp(-\lambda t)\}\,(1+\nu/\lambda)]G}{1-\theta\,[1-\exp(-\lambda t)]}\ .\text{ Hence}$$

$$\left(\frac{\partial G}{\partial \theta}\right)_{\theta=1} = 1+(1+\nu/\lambda)\,[\exp(\lambda t)-1]\,.\text{ When }\lambda = \nu,$$

$G = \theta\,\exp(-2\lambda t)\,\{1-\theta\,[1-\exp(-\lambda t)]\}^{-2}$. Coefficient of θ^n is $n\exp(-2\lambda t)$ $[1-\exp(-\lambda t)]^{n-1}$.

CHAPTER 3

Queueing Theory

3.1 INTRODUCTION

We have all had to wait for service. Often we are obliged to stand in a line, usually called a queue in the United Kingdom and a waiting line in the United States. The commoner experiences include waiting for public transport, obtaining entrance to a place of entertainment, and shopping at a supermarket. It is not surprising that the terminology used in the study of such processes should employ such words as *customer, server, traffic, waiting time,* and so forth. Note that waiting time may, or may not, include service time, and that the queue may, or may not, include the customer(s) being served.

There are situations where the queue, or assembly of waiting customers, is not visible to its individual members. This is the case of patients waiting for a hospital bed and of callers attempting to communicate through a telephone switchboard. Yet the same elements are present, namely demands for service and a facility for satisfying these demands.

The urge to study such systems is prompted by two obvious features. Owing to the ebb and flow of customers there will be some occasions when the service facility is not fully employed, and others where it is under continuous pressure, or *congested.* Indeed some authors suggest that 'congestion theory' rather than 'queueing theory' is a more appropriate description of the mathematical framework which has been erected. Costs are involved when the service is under-employed (low productivity), and in the congested regime (loss of productive time for queue members). It is, accordingly, one of the tasks of queueing theory to examine how these costs may be reduced by modifications to the mechanics of the system. In order to do this it is essential to postulate theoretical models from which primary measures of performance can be calculated. The model must define unambiguously at least the following elements:

(a) *The arrival process.* This is usually given in terms of the distribution of the time intervals between consecutive customer arrivals (*inter-arrival intervals*). A customer who can not obtain service immediately upon arrival is said to be **blocked**. A blocked customer can be said to be **held**, or

lost, accordingly as he does or does not wait for service. If the waiting room has only finite capacity, an arriving customer who finds it full is automatically lost. We shall define the number of customers in the queue to be *all* those present, either waiting or receiving service. This number, normally a random variable, is also called the *state* of the queue or *system*.

(b) *The service mechanism*. This is described by the number of servers and the distribution of the service time per customer. Servers are either **busy** or **idle** in the obvious sense, and it is generally assumed that an idle server is instantly available to a customer demanding service.

(c) *The queue discipline*. This includes the rules for determining the choice of next to be served from among those waiting. A common discipline is to take customers in order of arrival (**first come, first served**), though an appointment system may be equally 'fair'. Many variations are possible, including random selection for service (as in boarding a train), and the imposition of a priority rule (as in allocating public housing according to 'need'). If a new arrival finds servers idle it is usually assumed that he is allocated to one of these at random.

We next discuss the simplest queueing system. This will provide a background for more complex cases. Suppose that there is a single server who is always available to the next customer demanding service. The queue discipline is 'first come, first served' so that we picture customers waiting in a single line in order of arrival and there is unlimited capacity. We further suppose that, regardless of whether the server is busy or free, the times between customer arrivals have independent exponential distributions — each with parameter λ. Thus in an interval of duration t the number of arrivals has a Poisson distribution with parameter λt. In an interval of small duration $(t, t+\delta t)$, the probability of just one arrival is $\lambda\delta t + o(\delta t)$ and of no arrival is $1 - \lambda\delta t + o(\delta t)$. Regardless of the number of customers present, service times for individual customers are also to have independent exponential distributions but with parameter μ. This implies that if the server is busy at time t then the residual time to completion is again exponential with parameter μ and the probability that service will terminate in the interval $(t, t+\delta t)$ is $\mu\delta t + o(\delta t)$. If the server is idle, then of course no service can terminate and it is this aspect which causes the exact solution to be nontrivial. The probability of more than one event, either service completion or customer arrival in $(t, t+\delta t)$ is neglible in comparison with that of a single event, i.e. it is $o(\delta t)$.

3.2 DERIVING THE STATE EQUATIONS

Let $p_{mn}(t)$ be the probability that there are n customers in the queue (including the one being served if there be such) at time t, given that there were initially m customers. Then, for $n \geqslant 2$, there will be n present at time $t+\delta t$ if

(a)   there were n present at time t, and in the further interval $(t, t+\delta t)$ the
       service in progress does not terminate and no new customer arrives; the
       probability of this event is

$$p_{mn}(t) \ [1-\lambda\delta t+o(\delta t)] \ [1-\mu\delta t+o(\delta t)]$$
$$= p_{mn}(t) \ [1-(\lambda+\mu)\delta t] \ + o(\delta t).$$

(b)   there were $n+1$ present at time t, and in the interval $(t, t+\delta t)$ one service
       terminates and no fresh customer arrives; this event has probability

$$p_{m,n+1}(t) \ [1-\lambda\delta t+o(\delta t)] \ [\mu\delta t+o(\delta t)]$$
$$= p_{m,n+1}(t) \, \mu\delta t + o(\delta t);$$

(c)   there were $n-1 (\geqslant 1)$ customers at time t, and the service in progress does
       not terminate, but one new customer joins the queue, in the interval
       $(t, t+\delta t)$. The corresponding probability is

$$p_{m,n-1}(t) \ [\lambda\delta t+o(\delta t)] \ [1-\mu\delta t+o(\delta t)]$$
$$= p_{m,n-1}(t) \, \lambda\delta t + o(\delta t).$$

These are the only combinations that need to be taken into account. All
others entail the simultaneous occurrence of two or more events whose probability is $o(\delta t)$ as stated at the end of section 3.1. For example, if there are n in
the queue at time t, one could argue that there would again be n at time $t+\delta t$ if
during $(t, t+\delta t)$ both an arrival and a service termination occurred. But the
probability of this event is $[\lambda\delta t+o(\delta t)] \ [\mu\delta t+o(\delta t)]$ which is $o(\delta t)$.

Since the cases, (a), (b), (c), are mutually exclusive,

$$p_{mn}(t+\delta t) = [1-(\lambda+\mu)\delta t] \, p_{mn}(t) + \mu\delta t p_{m,n+1}(t) + \lambda\delta t p_{m,n-1}(t) + o(\delta t),$$

that is

$$\frac{p_{mn}(t+\delta t)-p_{mn}(t)}{\delta t} = -(\lambda+\mu)p_{mn}(t) + \mu p_{m,n+1}(t) + \lambda p_{m,n-1}(t) + \frac{o(\delta t)}{\delta t}.$$

Now let $\delta t \to 0$

$$\frac{dp_{mn}(t)}{dt} = -(\lambda+\mu)p_{mn}(t) + \mu p_{m,n+1}(t) + \lambda p_{m,n-1}(t), n \geqslant 2. \qquad (3.1)$$

We next consider the case $n = 1$.

$$p_{m1}(t+\delta t) = [1-(\lambda+\mu)\delta t] \, p_{m1}(t) + \mu\delta t p_{m2}(t) + \lambda\delta t p_{m0}(t) + o(\delta t)$$

leading to

$$\frac{dp_{m1}(t)}{dt} = -(\lambda+\mu)p_{m1}(t) + \mu p_{m2}(t) + \lambda p_{m0}(t), \qquad (3.2)$$

and this is the same as equation (3.1) when $n = 1$. The fact that, when the queue is empty, the probability that no service terminates (no-one leaves the queue) is 1 instead of $1 - \mu \delta t$ makes no difference as $\delta t \to 0$. However, for $n = 0$,

$$p_{m0}(t+\delta t) = 1.(1-\lambda \delta t)p_{m0}(t) + \mu \delta t p_{m1}(t) + o(\delta t),$$

since event (c) does not exist for $n = 0$, and 1 replaces $1 - \mu \delta t$ for event (a), and we have

$$\frac{dp_{m0}(t)}{dt} = - \lambda p_{m0}(t) + \mu p_{m1}(t). \tag{3.3}$$

This is not the same as (3.1) with $n = 0$, even after setting $p_{m,-1}(t) \equiv 0$.

The state equations (3.1), (3.2) and (3.3), together with $\sum_{n=0}^{\infty} p_{mn}(t) = 1$, $p_{mm}(0) = 1$, $p_{mn}(0) = 0$ $n \neq m$, are a full description of the queueing situation stated in 3.1.

3.3 SOLVING THE EQUATIONS

An exact solution, using the method of Laplace transforms, is to be found in more advanced discussions and involves Bessel functions. Rather surprisingly, a direct construction is available but the argument involves some sophisticated combinatorial results. (For both approaches see B. Conolly [1].) In the next problem we state the result for $m = 0$ and content ourselves with asking for it to be verified.

Problem 1. Let

$$I_n(y) = \sum_{j=0}^{\infty} \frac{(y/2)^{n+2j}}{j! \, (n+j)!}$$

and

$$\phi_n(t) = \exp \left[-(\lambda+\mu)t\right] (\lambda/\mu)^{n/2} I_n \left[2t\sqrt{(\lambda\mu)}\right]$$

then

$$p_{0n}(t) = \sum_{k=0}^{\infty} \left[\left(\frac{\lambda}{\mu}\right)^{-k} \phi_{n+k}(t) - \left(\frac{\lambda}{\mu}\right)^{-k-1} \phi_{n+k+2}(t) \right].$$

Verify that $p_{0n}(t)$ satisfies equation (3.1) for $n \geqslant 1$. It is easily shown, but you may assume, that

$$\frac{d}{dy} \left[I_n(y)\right] = \frac{1}{2} \left[I_{n-1}(y) + I_{n+1}(y)\right]. \quad \blacksquare$$

Problem 2. We have found the forward equations. Derive the following backward equations for the simple single server queue.

$$\frac{dp_{1n}(t)}{dt} = \mu p_{0n}(t) - (\lambda+\mu)p_{1n}(t) + \lambda p_{2n}(t),$$

$$\frac{dp_{0n}(t)}{dt} = \lambda p_{1n}(t) - \lambda p_{0n}(t). \quad \blacksquare$$

Problem 3. Let,

$$\frac{dp_n(t)}{dt} = \rho p_{n-1}(t) - (1+\rho)p_n(t) + p_{n+1}(t), n \geqslant 1$$

and

$$\frac{dp_0(t)}{dt} = -\rho p_0(t) + p_1(t),$$

where

$$\sum_{n=0}^{\infty} p_n(t) = 1, 0 \leqslant p_n(t) \leqslant 1, 0 \leqslant n \leqslant \infty.$$

Show that if $G(\theta,t) = \sum_{n=0}^{\infty} p_n(t)\,\theta^n$, then

$$\theta\frac{\partial G}{\partial t} = (\theta-1)\ [(\rho\theta-1)G + p_0(t)].$$

Hence or otherwise show that $E(N) = \sum_{1}^{\infty} np_n(t)$ satisfies $\dfrac{dE(N)}{dt} = p_0(t) + \rho - 1$.
(Hint: differentiate both sides with respect to θ and then let $\theta \to 1$.)
    What do you observe if $p_0(t) \to 1-\rho$ as $t \to \infty$? $\blacksquare$

Problem 4. Suppose for the simple single server queue that we impose the severe restriction that an arriving customer who finds the server busy is turned away. This is called a loss system. That is, only $p_{00}(t), p_{01}(t), p_{10}(t), p_{11}(t)$ are non-zero and $p_{10}(t) + p_{11}(t) = 1, p_{00}(t) + p_{01}(t) = 1$. Show that

$$\frac{dp_{10}(t)}{dt} = -\lambda p_{10}(t) + \mu p_{11}(t) \text{ and hence that}$$

$$\frac{d}{dt}\{\exp[(\lambda+\mu)t]\,p_{10}(t)\} = \mu \exp\ [(\lambda+\mu)t].$$

Deduce that

$$p_{10}(t) = \frac{\mu}{\lambda+\mu}\ \{1-\exp\ [-(\lambda+\mu)t]\}$$

and state the value of $\lim_{t\to\infty}\ [p_{10}(t)].$ $\blacksquare$

3.4 EQUILIBRIUM DISTRIBUTION

Although the exact equations are so intractable, matters are much simplified if we assume that, as t increases, $p_{mn}(t)$ tends to a limit, p_{mn}, which is independent of t. In that case, $\dfrac{\mathrm{d}p_{mn}(t)}{\mathrm{d}t}$ tends to zero and the *equilibrium equations*, thus obtained from (3.1) are

$$0 = \mu p_{m,n+1} - (\lambda+\mu)p_{mn} + \lambda p_{m,n-1} \quad n \geqslant 1. \tag{3.4}$$

These are readily solved, for we can re-arrange (3.4) as

$$\mu(p_{m,n+1} - p_{mn}) = \lambda(p_{mn} - p_{m,n-1})$$

and sum over n from 1 to $r-1$. Thus

$$\mu \sum_{n=1}^{r-1} (p_{m,n+1} - p_{mn}) = \lambda \sum_{n=1}^{r-1} (p_{mn} - p_{m,n-1})$$

and, profiting from the fact that the second suffix is 'out of step' by 1 on each side, arrive at

$$\mu(p_{mr} - p_{m1}) = \lambda(p_{m,r-1} - p_{m0}).$$

But from the equilibrium equation for $n = 0$, obtained from (3.3),

$$0 = \mu p_{m1} - \lambda p_{m0}$$

and finally

$$\mu p_{mr} = \lambda p_{m,r-1}, \quad r \geqslant 1,$$

or

$$p_{mr} = \frac{\lambda}{\mu}\, p_{m,r-1}, \text{ and by continued application}$$

$$p_{mr} = \left(\frac{\lambda}{\mu}\right)^r p_{m0}. \tag{3.5}$$

We pause to observe that if $(\lambda/\mu) > 1$, then eventually $p_{mr} > 1$ for some least r unless $p_{m0} = 0$. But $p_{m0} = 0$ forces $p_{mr} = 0$ for every r. Thus (3.5) only makes sense if $\dfrac{\lambda}{\mu} < 1$ and this must be a necessary condition for the limiting probabilities to exist. This requirement agrees with the intuitive idea that unless the mean service time is less than the mean time between arrivals, then eventually the queue must pile up beyond all bounds. The ratio $\rho = \lambda/\mu$ is known as the **traffic intensity**. If then $\rho < 1$, we sum (3.5) over r

$$\sum_{r=0}^{\infty} p_{mr} = 1 = p_{m0} \sum_{r=0}^{\infty} (\lambda/\mu)^r$$
$$= p_{m0} \left[1/(1-\lambda/\mu)\right].$$

That is to say

$$p_{m0} = 1-\rho \text{ and } p_{mr} = (1-\rho)\rho^r, \ r \geqslant 0.$$

We see, possibly belatedly, that if p_{mr} exists it does not depend on m at all and we may write p_r. The p.d.f. $(1-\rho)\rho^r$ is that of a geometric distribution with parameter $(1-\rho)$, and we can readily calculate several measures which are relevant to the performance of the queue after it has been running for some long time.

(a)   The probability that the queue contains at least k customers is

$$\sum_{r=k}^{\infty} (1-\rho)\rho^r = \rho^k;$$

(b)   The expected number in the queue is

$$E(N) = \sum_{1}^{\infty} np_n = \sum_{1}^{\infty} n(1-\rho)\rho^n$$

$$= \rho(1-\rho) \sum_{1}^{\infty} n\rho^{n-1} = \rho(1-\rho)\frac{1}{(1-\rho)^2} = \frac{\rho}{1-\rho}.$$

The reader should verify that $V(N) = \rho/(1-\rho)^2$.

We have omitted to discuss the behaviour of the queue when $\rho = 1$. The results (a) and (b) helps us to see why this case must be excluded. For if we allow ρ to approach 1, by (a), the probability of at least k customers present tends to 1 for every k. Further, from (b), as ρ approaches 1, the expected number of customers becomes unbounded. In fact the whole argument of this section would break down when $\rho = 1$. For in the equilibrium equations, if they exist, this would force $p_0 = p_1 = \ldots$. But this is not possible unless all these probabilities are 0! We now continue with our discussion of the queue in equilibrium.

(c)   The limiting probability that the server is idle is p_0. From the frequency point of view this implies that if several such queues are inspected, then a proportion p_0 of the servers are expected to be idle. For a single queue, p_0 may also be taken as the proportion of time that its server is idle (in the limit).

(d)   The probability that the queue contains n customers, given that it is not empty, is

$$Pr(N=n \mid N>0) = \frac{Pr(N=n, n>0)}{Pr(N>0)} = \frac{(1-\rho)\rho^n}{\rho} = (1-\rho)\rho^{n-1}.$$

Hence

$$E(N|N>0) = \sum_1^\infty n(1-\rho)\rho^{n-1} = (1-\rho) \sum_1^\infty n\rho^{n-1}$$

$$= (1-\rho)/(1-\rho)^2 = 1/(1-\rho) = \mu/(\mu-\lambda).$$

(e)   The interval between the end of one idle period and the commencement of the next idle period is called a **busy** period. Approximately, in a long time t, the queue is empty for a total time $p_0 t$ and non-empty for a time $(1-p_0)t$. Since the mean duration of an idle period is $1/\lambda$, the number of such periods is $\lambda p_0 t$. Since busy periods alternate with idle periods, the expected duration of a busy period is $(1-p_0)t/\lambda p_0 t = 1/(\mu-\lambda)$.

3.5 WAITING TIME

We next look at the simple, single server queue from the point of view of a particular customer. He is certainly concerned with his **waiting time** — the duration from the moment of his arrival until his departure, W. If on joining the queue he finds n in front, then he must endure the sum of $(n+1)$ independent exponential service times, which has the $\Gamma(n+1,\mu)$ distribution, with expected waiting time $(n+1)/\mu$, regardless of the time already spent by any customer then at the service point. Also,

$$Pr(W \leqslant t|n) = \int_0^t \frac{\mu(\mu x)^n \exp(-\mu x)}{n!} \, dx.$$

Before he arrives he faces a probability $p_{mn}(t)$ of finding n customers in front. We must undertake a brief discussion of this last point. An arriving customer does not join the queue at a fixed time. So how do we know that the probability of finding n in front at an arrival time is the same as finding that number in the system at the same time regarded as fixed? That this is not a property to be lightly assumed will emerge from the following example. Suppose customers arrive at regular fixed intervals t_0 and that services are of fixed duration $t_0/2$. Then every arriving customer, if the system is initially empty, is bound to find no-one in front. But an outside observer will see one customer for the duration of service and then no-one until the next arrival. This outside observer will, over a long interval, find that the queue is empty for about half the time. In fact, the equivalence of the two distributions in question is a consequence of the Poisson arrival rate. This is 'almost obvious' since at a fixed time t, the probability of an arrival in $(t,t+\delta t)$ is $\lambda \delta t + o(\delta t)$ whatever the time and however many there are in the queue at time t. Further confirmation is provided in the next (optional) problem.

Problem 5. The inter-arrival times for customers have independent exponential distributions with parameter λ. T_n is the arrival time of the nth customer. Show that the conditional p.d.f. of T_1 given T_n is $(n-1)(t_n-t_1)^{n-2}/t_n^{n-1}$ for $0 < t_1 < t_n$ $(n \geqslant 2)$. Show that this is also the conditional p.d.f. of T_1 given that there are just $n-1$ arrivals in the *fixed* interval $(0, t_n)$. Hence the probability that the first customer is still present at time t_n is the same in either case and depends only on the service distribution which has not been specified. ∎

If the arriving customer joins sufficiently late in the day, we may take the probability of finding n in front as the equilibrium probability $p_n = (1-\rho)\rho^n$. The unconditional expected waiting time is then

$$\sum_{n=0}^{\infty} (n+1)p_n/\mu = \left[\sum_0^{\infty} np_n + \sum_0^{\infty} p_n \right]/\mu$$

$$= (\frac{\rho}{1-\rho} + 1)/\mu$$

$$= \frac{1}{\mu-\lambda}.$$

On the same basis, the unconditional probability that the waiting time is less than t is

$$\sum_{n=0}^{\infty} Pr(W \leqslant t \mid n)p_n$$

$$= \sum_0^{\infty} \left\{ \left[\int_0^t \frac{\mu(\mu x)^n \exp(-\mu x)}{n!} dx \right] (1-\rho)\rho^n \right\}$$

$$=$$

$$= \int_0^t \left[\mu(1-\rho)\exp(-\mu x) \sum_0^{\infty} \frac{(\mu x\rho)^n}{n!} \right] dx$$

$$= \int_0^t \mu(1-\rho)\exp(-\mu x)\exp(\mu\rho x) dx$$

$$= \int_0^t \mu(1-\rho)\exp[-\mu(1-\rho)x] dx.$$

The integrand is the p.d.f. of an exponential distribution with parameter $\mu(1-\rho)$.

The subject of the next problem concerns the closely related question of an arriving customer's **queueing time**, which lasts from the moment of his arrival until his service begins.

Problem 6. The queueing time has a mixed distribution, for there is a probability $p_{m0}(t)$ of finding the queue empty so that the queueing time is zero. By an argument similar to the one involving the waiting time, show that, in the equilibrium distribution, the queueing time has an exponential distribution with parameter $\mu(1-\rho)$, *conditional* on it being positive. ∎

3.6 MODIFICATIONS OF THE SIMPLE QUEUE

Many points now clamour for attention. What happens if there is more than one server? What is to be done if the service time cannot safely be taken as exponential? What is the effect of some customers having priority? We evidently need a compact notation for describing the main features of a queue.

(a)   If the inter-arrival times have a general distribution, this is denoted by GI.
(b)   If the service times have a general distribution, this is denoted by G.
(c)   If the distributions in (a) or (b) happen to be exponential, gamma (index n) or constant then the letters M, E_n, D respectively are employed.
(d)   The number of servers is denoted by s.

Unless otherwise stated, the times in (a), (b) are taken to be independent.

The system is then described by the appropriate letter in the order: arrival distribution/service distribution/number of servers. For example, $D/M/2$ states that the customers arrive at regular intervals, the service time is exponential and that there are two servers. The simple queue we have examined so closely is of course $M/M/1$.

When there are n customers in the queue, we shall assume that the probability of a new arrival in the interval $(t, t+\delta t)$ is $\lambda_n \delta t + o(\delta t)$, regardless of t. For example, for the $M/M/1$ queue, if the waiting room is strictly limited to no more than m, then $\lambda_n = \lambda$, $n < m$, but $\lambda_n = 0$, $n \geqslant m$. While if a service is in progress the corresponding probability that it terminates in the interval $(t, t+\delta t)$ is $\mu_n \delta t + o(\delta t)$. For example, suppose there are s independent servers having exponential service times with parameter μ and some of these are busy. Since any one of these may terminate in $(t, t+\delta t)$ it is appropriate to take $\mu_n = n\mu(n \leqslant s)$ but $\mu_n = s\mu(n \geqslant s)$. The parameters λ_n, μ_n are thus state-dependent but not time-dependent. It is tempting to suppose that if a service starts when there are n in the queue then its duration has an exponential distribution with parameter μ_n. Unfortunately, if in the course of a service additional customers arrive the server adjusts his rate.

Except in special cases, we have also lost the Poisson distribution for the number of arrivals in a fixed time. Using the argument of paragraph 3.2 we can still derive the forward equations for $p_{mn}(t)$. These are

$$\frac{dp_{mn}(t)}{dt} = \mu_{n+1}p_{m,n+1}(t) - (\mu_n + \lambda_n)p_{mn}(t) + \lambda_{n-1}p_{m,n-1}(t), n \geqslant 1;$$

$$\frac{dp_{m0}(t)}{dt} = \mu_1 p_{m1}(t) - \lambda_0 p_{m0}(t).$$

(3.6)

If we suppose that $p_{mn} = \lim\limits_{t\to\infty} p_{mn}(t)$ exists, the equilibrium equations are

$$\mu_{n+1}p_{m,n+1} - \mu_n p_{mn} = \lambda_n p_{mn} - \lambda_{n-1}p_{m,n-1}, n \geqslant 1; \qquad (3.7)$$

$$\mu_1 p_{m1} = \lambda_0 p_{m0}. \qquad (3.8)$$

Summing equation (3.7) from 1 to $r-1$,

$$\mu_r p_{mr} - \mu_1 p_{m1} = \lambda_{r-1}p_{m,r-1} - \lambda_0 p_{m0}$$

or, in virtue of (3.8),

$$p_{mr} = \frac{\lambda_{r-1}}{\mu_r}\, p_{m,r-1}, r \geqslant 1,$$

and by continued application

$$p_{mn} = \prod_{r=1}^{n} \frac{\lambda_{r-1}}{\mu_r} p_{m0}. \qquad (3.9)$$

But provided that $\sum\limits_{0}^{\infty} p_{mn} = 1$, we have, by summing (3.9) from 1 to ∞,

$$1 - p_{m0} = \sum_{n=1}^{\infty} \prod_{r=1}^{n} \frac{\lambda_{r-1}}{\mu_r} p_{m0},$$

$$p_{m0} = 1 / \left[1 + \sum_{n=1}^{\infty} \prod_{r=1}^{n} \frac{\lambda_{r-1}}{\mu_r} \right]. \qquad (3.10)$$

If $p_{m0} \neq 0$, the equality in (3.10) requires the convergence of $\sum\limits_{n=1}^{\infty} \prod\limits_{r=1}^{n} (\lambda_{r-1}/\mu_r)$ and this must be a necessary condition for the existence of limiting probabilities, for otherwise by (3.9) $p_{mn} = 0$ for every finite n. It is now clear that p_{mn} does not depend on m and we may write merely p_n.

Example 1. M/M/s queue
Here we have random arrivals $\lambda_n = \lambda \ (n \geqslant 0)$. If $r(\leqslant s)$ servers are busy, then the distribution of time to the next service completion is the distribution of the minimum of r service times, each exponentially distributed with parameter μ, and this is another exponential distribution with parameter $r\mu$. We take $\mu_r = r\mu \ (r \leqslant s), \mu_r = s\mu, (r \geqslant s)$.

$$\prod_{r=1}^{n} \frac{\lambda_{r-1}}{\mu_r} = \frac{\lambda^n}{(s\mu)^{n-s}\mu^s s!} \quad \text{for } n \geqslant s$$

$$\prod_{r=1}^{n} \frac{\lambda_{r-1}}{\mu_r} = \frac{\lambda^n}{\mu^n n!} \quad \text{for } 1 \leqslant n < s.$$

Hence

$$p_0 = \left[1 + \sum_{n=1}^{s-1} \frac{\lambda^n}{\mu^n n!} + \sum_{n=s}^{\infty} \frac{\lambda^n}{(s\mu)^{n-s}\mu^s s!} \right]^{-1} \tag{3.11}$$

Since the first sum has only a finite number of terms we are only concerned, in establishing $p_0 \neq 0$, with

$$\sum_{n=s}^{\infty} \frac{\lambda^n}{(s\mu)^{n-s}\mu^s s!} = \frac{1}{s!} \left(\frac{\lambda}{\mu}\right)^s \sum_{n=s}^{\infty} \left(\frac{\lambda}{s\mu}\right)^{n-s}.$$

Clearly $\sum_{n=s}^{\infty} \left(\frac{\lambda}{s\mu}\right)^{n-s}$ converges if and only if $\lambda/s\mu < 1$ which confirms that the behaviour of the queue in the long run is determined by its ability to cope when all the servers are busy.

To summarise:

$$p_n = \frac{s^s}{s!} \cdot \left(\frac{\lambda}{s\mu}\right)^n p_0, \quad n \geqslant s, \tag{3.12}$$

$$p_n = \frac{1}{n!} \left(\frac{\lambda}{\mu}\right)^n p_0, \quad 0 \leqslant n < s,$$

where p_0 is calculated from (3.11).

Problem 7. For the $M/M/s$ queue show that, if the equilibrium distribution exists, $Pr(N=n \mid N \geqslant s) = (1-\lambda/s\mu)(\lambda/s\mu)^{n-s}, n = s, s+1, \ldots$ ∎

Problem 8. For the $M/M/2$ queue show that, if the equilibrium distribution exists and $\rho = \lambda/2\mu$, then $p_0 = (1-\rho)/(1+\rho)$, $p_n = 2\rho^n p_0$ and the expected number in the queue is $2\rho/(1-\rho^2)$. We may compare these results with the corresponding $M/M/1$ queue with a view to deciding whether the extra server is worthwhile. The additional expense must be justified in terms of gains in other directions. ∎

Problem 9. For the $M/M/s$ queue in the limiting distribution, examine the situation as the number of servers increases without limit. Show that $p_0 \to \exp(-\lambda/\mu)$ as $s \to \infty$ and $p_n \to (\lambda/\mu)^n \exp(-\lambda/\mu)/n!$, i.e. corresponds to a Poisson distribution with parameter λ/μ. ∎

Problem 10. If in a modified queue, a long waiting line discourages arrivals, suppose that $\lambda_n = \lambda/(n+1)$, $\mu_n = \mu$. Show that in the equilibrium distribution, $p_0 = \exp(-\lambda/\mu)$ and $p_n = (\lambda/\mu)^n \exp(-\lambda/\mu)/n!$ ∎

Problem 11. If in a modified queue, $\mu_n = \mu$ but there is waiting room for at most m customers we could take $\lambda_n = \lambda(n < m)$, $\lambda_n = 0$, $(n \geqslant m)$. Show that $p_0 = 1/\sum_{n=0}^{m} (\lambda/\mu)^n$ and calculate p_n. ∎

3.7 FURTHER REMARKS ON THE $M/M/s$ QUEUE

We compute some other measures for the equilibrium distribution. We first find the expected number of idle servers. If there are i in the queue then $s - i$ servers are idle $(i \leqslant s-1)$, and none is idle for $i \geqslant s$. Hence the required expectation is

$$\sum_{i=0}^{s-1} (s-i)p_i = s \sum_{i=0}^{s-1} p_i - \sum_{i=0}^{s-1} ip_i,$$

$$= s(1 - \sum_{i=s}^{\infty} p_i) - \sum_{i=1}^{s-1} ip_i. \qquad (3.13)$$

Now from (3.7), (3.8) we deduced that, in the equilibrium distribution, $p_i = (\lambda_{i-1}/\mu_i)p_{i-1}$. For the $M/M/s$ queue, we have $p_i = (\lambda/i\mu)p_{i-1}$ if $i \leqslant s$ and $p_i = (\lambda/s\mu)p_{i-1}$, if $i \geqslant s$. Substituting these values in (3.13), the expected number of idle servers is

$$s - s \sum_{i=s}^{\infty} \left(\frac{\lambda}{s\mu}\right)p_{i-1} - \sum_{i=1}^{s-1} \left(\frac{i\lambda}{i\mu}\right)p_{i-1}$$

$$= s - \frac{\lambda}{\mu}\left(\sum_{i=s}^{\infty} p_{i-1} + \sum_{i=1}^{s-1} p_{i-1}\right)$$

$$= s - (\lambda/\mu) = s(1-\rho), \text{ where } \rho = \lambda/s\mu. \qquad (3.14)$$

E (number of idle servers) $= s(1-\rho)$. Whence of course
E (number of active servers) $= s-s(1-\rho) = s\rho$ or
E (proportion of active servers) $= \rho$. $\qquad (3.15)$

The result throws light on why the traffic density ρ is also called the **server utilisation** (see section 3.15).

Problem 12. If it is given that only r $(\leqslant s)$ of the servers are active, show that the conditional probability that a particular server is idle is $(s-r)/s$. Hence or otherwise show that the unconditional probability that a particular server is idle is $1-\lambda/\mu s$. (Hint: Use equation (3.14).) ∎

Waiting Time (M/M/s)
In the equilibrium situation, if an arriving customer finds fewer than s customers in front he does not have to queue at all to start service. If he finds $n(\geqslant s)$ in front he must queue until the end of $n-s+1$ services. While all s servers are busy, the time to termination of some service has the exponential distribution with parameter $s\mu$. Thus the average duration of each of these services is $1/s\mu$. Hence the mean queueing time is

$$\sum_{n=s}^{\infty} \frac{n-s+1}{s\mu} p_n,$$

using (3.12)

$$= \sum_{n=s}^{\infty} \left(\frac{n-s+1}{s\mu}\right) \frac{s^s}{s!} \rho^n p_0$$

$$= \frac{(s\rho)^s p_0}{s!(s\mu)} \sum_{n=s}^{\infty} (n-s+1)\rho^{n-s}$$

$$= \frac{(s\rho)^s}{s!(s\mu)} (1-\rho)^{-2} p_0. \tag{3.16}$$

For example, when $s = 2$, $\rho = \lambda/(2\mu)$ and, from problem 8, $p_0 = (1-\rho)/(1+\rho)$, hence the expected queueing time, from (3.16), is $\rho^2/[\mu(1-\rho^2)]$.

For the expected waiting time, we must add to the expecting queueing time the mean service of the arriving customer, which is $1/\mu$. Thus the expected waiting time when $s = 2$ is $1/[\mu(1-\rho^2)]$.

3.8 MEASURES OF PERFORMANCE

Example 2
Suppose there are two types of customer, type I arriving in a Poisson stream at rate λ_1 and type II in a Poisson stream at rate λ_2. The service time of each of two independent servers is exponential with mean $1/\mu$. Now one server could be allocated entirely to type I customers and the other to type II customers. In that case we have two independent $M/M/1$ queues and if equilibrium distributions exist for both, with traffic intensities ρ_1, ρ_2, then the mean lengths of the individual queues are $\rho_1/(1-\rho_1)$, $\rho_2/(1-\rho_2)$ respectively. Now it is apparent that there is some loss of server utilisation because there will be occasions when one server is idle and yet the other has a customer waiting. We can cure this defect by making all arriving customers wait in one line and take the first available server, i.e. convert to an $M/M/2$ queue. The arrival rate is $\lambda_1 + \lambda_2$ and the traffic intensity $\rho = (\lambda_1 + \lambda_2)/2\mu = (\rho_1 + \rho_2)/2$, and the mean number in the queue is now $2\rho/(1-\rho^2)$. For instance if $\rho_1 = \frac{1}{5}$, $\rho_2 = \frac{4}{5}$, then $\rho = \frac{1}{2}$. The

corresponding mean numbers are $\frac{1}{4}, 4, \frac{4}{3}$. The service manager is pleased with the re-arrangement. The mean waiting time over all customers is $1/[\mu(1-\rho^2)] = 4/3\mu$. Type I customers are possibly not so happy since with their own private queue the mean waiting time was $1/[\mu(1-\rho_1)] = 5/4\mu$. Indeed they might lay claims to 'priority' and ask that if a server becomes free he takes the earliest type I customer next.

The last example draws attention to the fact that it is far from clear what is a 'best' arrangement for a queueing process. Even for the simplest $M/M/1$ queue there are several plausible measures of performance. For instance,

(a)   the probability that the server is free,
(b)   the expected number in the queue,
(c)   the probability that m or more are in the queue,
(d)   the average queueing time,
(e)   the average waiting time.

But these measures are inter-related. If the traffic intensity decreases, then (b) decreases but (a) increases!

There are various ways in which the running of a queueing system may be modified with a view to improving its performance. In certain cases, we can intervene in the arrival rate: for example, we can

(1)   allocate blocks of time for particular groups of customers,
(2)   impose waiting room restrictions,
(3)   arrange for fixed appointment times.

We might also impose special rules to decide the service order after arrival. For instance, priority can be given to certain kinds of customer or it can be ensured that, on arrival, he has access to a 'fast lane' if his demand be small. We can always improve the service rate by increasing the number of servers. Whether it is possible to alter the service distribution itself seems a more open matter. Whether service is provided by machines or humans there will be limits to the extent to which service can be speeded up. Moreover the length of service may be determined by the demand of the customer.

3.9 GENERAL SERVICE DISTRIBUTION (SINGLE SERVER)

If we no longer assume that service times are exponential then we lose one of the supports on which our previous calculations rested, namely that if we look at the queue at any time and find a customer being served, then the residual service time for that customer is again exponential. To avoid this difficulty we consider the state of the queue only at those moments when a customer departs, when the residual service is of course zero. The number left behind by such a departing customer is the number in the queue *at that moment*. We find a relation between the probabilities that stated numbers were left behind by successive customers in the queue. If customer number $m - 1$ left no-one behind, then the number left behind by customer number m is just the number that arrived behind him during the course of *his* service. If customer number $m - 1$ left $i(> 0)$ behind him, then customer number m is the first of these and he already has $i - 1$ behind him as his service starts. Thus if customer number m leaves j behind him, then $j - (i-1)$ arrivals took place during his service time. But the number of arrivals must be non-negative, hence $j \geqslant i - 1$ or $i \leqslant j + 1$ ($i > 0$). Let $p_j^{(m)}$ be the probability that customer m leaves j behind and let p_{ij} be the conditional probability that *any* departing customer leaves j behind given that the previous customer left i behind him. Then evidently $p_{ij} = 0$ if $j < (i - 1)$ for $i > 0$. Now customer m left j if customer $m - 1$ left i and subsequently the next customer left j for any permitted i. Hence

$$p_j^{(m)} = \sum_{i=0}^{j+1} p_i^{(m-1)} p_{ij}. \qquad (3.17)$$

Suppose now that a limiting distribution exists for $p_j^{(m)}$ as m increases i.e. after sufficiently many customers have been served, then the number left by departing customers has settled down to an equilibrium distribution. If $p_j^{(m)} \to p_j$ then also $p_j^{(m-1)} \to p_j$ and, from (3.17)

$$p_j = \sum_{i=0}^{j+1} p_i p_{ij}. \qquad (3.18)$$

Now, from the discussion above, if $i > 0, p_{ij} = Pr(j-i+1)$ arrivals in one service) $= \pi_{j-i+1}$ say, and if $i = 0, p_{0j} = Pr(j$ arrivals in one service) $= \pi_j$. (3.18) is then

$$p_j = p_0 p_{0j} + \sum_{i=1}^{j+1} p_i p_{ij}$$
$$= p_0 \pi_j + \sum_{i=1}^{j+1} p_i \pi_{j-i+1}. \qquad (3.19)$$

Even supposing we can compute all the π_j, these equations are tiresome to solve. If $j = 0$,

$$p_0 = p_0 \pi_0 + p_1 \pi_0.$$

We express p_1 in terms of p_0 and now set $j = 1$ and find p_2 in terms of p_0 and so on. Eventually we use the condition $\sum\limits_{i=0}^{\infty} p_i = 1$ to obtain all the p_i. This is a situation where its easier to manipulate the corresponding generating functions,

$H(\theta) = \sum\limits_{j=0}^{\infty} \pi_j \theta^j$, $G(\theta) = \sum\limits_{j=0}^{\infty} p_j \theta^j$. To this end, multiply both sides of equation (3.19) by θ^j, where $|\theta| < 1$, and sum.

$$\sum_{j=0}^{\infty} p_j \theta^j = p_0 \sum_{j=0}^{\infty} \pi_j \theta^j + \sum_{j=0}^{\infty} \sum_{i=1}^{j+1} p_i \pi_{j-i+1} \theta^j$$

$$G(\theta) = p_0 H(\theta) + \sum_{i=1}^{\infty} \sum_{j=i-1}^{\infty} p_i \pi_{j-i+1} \theta^j, \tag{3.20}$$

after interchanging the order of summation. To simplify the double summation, we split θ^j to match the suffix of π_{j-i+1}.

Hence,

$$\sum_{i=1}^{\infty} \sum_{j=i-1}^{\infty} p_i \pi_{j-i+1} \theta^{j-i+1} \theta^{i-1} = \sum_{i=1}^{\infty} p_i \theta^{i-1} \left(\sum_{j=i-1}^{\infty} \pi_{j-i+1} \theta^{j-i+1} \right)$$

$$= \theta^{-1} \sum_{i=1}^{\infty} p_i \theta^i \left[H(\theta) \right]$$

$$= \theta^{-1} \left[G(\theta) - p_0 \right] H(\theta).$$

Thus (3.20) may be written

$$\theta G(\theta) = \theta p_0 H(\theta) + G(\theta) H(\theta) - p_0 H(\theta),$$

or

$$G(\theta) = p_0 H(\theta) (\theta - 1) / [\theta - H(\theta)]. \tag{3.21}$$

we have still to evaluate p_0. Setting $\theta = 0$ only produces $p_0 = p_0$. If we put $\theta = 1$ then $G(1) = 1$, but as $H(1) = 1$ both the numerator and the denominator vanish. To evaluate the limit as $\theta \to 1$ write (3.21) in the form

$$[\theta - H(\theta)] \, G(\theta) = p_0 H(\theta) (\theta - 1). \tag{3.22}$$

Now differentiate both sides of (3.22) with respect to θ and then set $\theta = 1$ to obtain

$$G(1) = p_0 / [1 - H'(1)].$$

Finally we have $p_0 = 1 - H'(1) = 1 - \Sigma j \pi_j = 1 - E(R)$, where R is the number of arrivals in one service. In principle, we can now find, in the limiting distribution, the expected number of customers left by a departing customer. This is

$\Sigma j p_j = G'$ (1). From (3.22) it can be shown, after differentiating twice with respect to θ, that

$$G'(1) = H'(1) + \frac{H''(1)}{2[1-H'(1)]} ,$$

(3.23)

$$= E(R) + \frac{E[R(R-1)]}{2[1-E(R)]}$$

(3.24)

3.10 GENERAL SERVICE, POISSON ARRIVALS

We now specialise the above discussion to the case when customers arrive in a Poisson stream at rate λ per unit time. Let the service time be a continuous random variable with p.d.f. $f(t)$. If the service lasts time t then the probability of r new arrivals is

$$\frac{(\lambda t)^r}{r!} e^{-\lambda t}.$$

Hence the unconditional probability of r arrivals in a complete service is

$$\pi_r = \int_0^\infty \frac{(\lambda t)^r}{r!} e^{-\lambda t} f(t) dt .$$

(3.25)

Furthermore, the expected number of arrivals in time t is λt, hence the unconditional expected number of arrivals in one complete service is

$$\int_0^\infty \lambda t f(t) dt = \lambda E(T)$$

(3.26)

(3.22) becomes $p_0 = 1 - E(R) = 1 - \lambda E(T)$ and a necessary condition for an equilibrium distribution is

$$\rho = \lambda E(T) < 1.$$

In the same way

$$E(R^2|t) = V(R|t) + E^2 (R|t)$$

$$= \lambda t + (\lambda t)^2 ,$$

$$E[R(R-1) = E(R^2) - E(R)$$

$$= \lambda^2 \int_0^\infty t^2 f(t) dt$$

$$= \lambda^2 [V(T) + E^2(T)] ,$$

and (3.24) becomes

$$G'(1) = \rho + \frac{\lambda^2 V(T) + \rho^2}{2(1-\rho)} .$$

(3.27)

Of formula (3.27) we may remark that, even if ρ is fixed, the expected number left by a departing customer can be decreased by 'choosing' a service distribution with a small variance. In this respect the constant service time has zero variance.

Example 3

Suppose the service times are exponential with parameter μ. Then $E(T) = 1/\mu$, $V(T) = 1/\mu^2$. Thus $\rho = \lambda/\mu$,

$$p_0 = 1 - \rho,$$

$$G'(1) = \rho + 2\rho^2/2(1-\rho) = \rho/(1-\rho).$$

The π_r can be calculated explicitly from (3.25).

$$\pi_r = \int_0^\infty \frac{(\lambda t)^r}{r!} e^{-\lambda t} \mu e^{-\mu t} dt$$

$$= \lambda^r \mu \int_0^\infty \frac{t^r}{r!} e^{-(\lambda+\mu)t} dt$$

$$= \frac{\mu \lambda^r}{(\lambda+\mu)^{r+1}} = \frac{1}{1+\rho} \left(1 - \frac{1}{1+\rho}\right)^r, \quad r = 0, 1, \ldots .$$

It is easily checked that R has mean ρ as anticipated.

$$H(\theta) = \sum_{r=0}^\infty \pi_r \theta^r = \frac{1}{1+\rho} \sum_{r=0}^\infty \left(\frac{\rho\theta}{1+\rho}\right)^r$$

$$= \frac{1}{1+\rho-\rho\theta}.$$

After substituting in (3.21), we easily obtain

$$G(\theta) = p_0/(1-\rho\theta) = \sum_{n=0}^\infty p_n \theta^n.$$

The coefficient of θ^n is $p_0 \rho^n = (1-\rho)\rho^n$. This is the equilibrium probability for n in the queue at any time for the system $M/M/1$ (not just at moments of departure).

Problem 13. $H(\theta)$ may sometimes be more conveniently computed by interchanging summation and integration, that is, sum $\sum_{r=0}^\infty \pi_r \theta^r$ within the integral form for π_r. Do this for Example 3. ■

Problem 14. In a single server queue, customers arrive in a Poisson process at rate λ per unit time and service times are independent and have p.d.f. $f(t) = \mu(\mu t) \exp(-\mu t)$. Show that in the limiting distribution,

$$G(\theta) = p_0 \mu^2 / [\mu^2 - 2\mu\lambda\theta + \lambda^2 \theta(\theta-1)], \ (\mu > 2\lambda),$$

where $G(\theta) = \sum\limits_{r=0}^{\infty} p_r \theta^r$ and p_r is the probability that a departing customer leaves r in the queue. Calculate p_0 and the mean number of customers left in the queue by a departing customer.

Example 4

Suppose the service time is constant, say t_0, and arrivals are in a Poisson stream at rate λ. This situation might apply if standard cargoes arrive at random and take a fixed time to unload. The probability of r arrivals in time t_0 is

$$(\lambda t_0)^r \exp(-\lambda t_0)/r!.$$

The probability generating function, $H(\theta)$, for the π_r is

$$\begin{aligned} H(\theta) &= \sum_{r=0}^{\infty} \pi_r \theta^r \\ &= \exp(-\lambda t_0) \sum_{r=0}^{\infty} \frac{(\lambda t_0 \theta)^r}{r!} \\ &= \exp(-\lambda t_0 + \lambda\theta t_0). \end{aligned}$$

From (3.21),

$$\begin{aligned} G(\theta) &= \frac{p_0 H(\theta)(\theta-1)}{\theta - H(\theta)} \\ &= \frac{p_0(\theta-1)\exp[-\lambda t_0(1-\theta)]}{\theta - \exp[-\lambda t_0(1-\theta)]} \end{aligned}$$

and

$$p_0 = 1 - E \text{ (number of arrivals in service of length } t_0)$$

$$= 1 - \lambda t_0 = 1 - \rho$$

which implies that $t_0 < 1/\lambda$. From (3.27)

$$\begin{aligned} G'(1) &= \rho + \rho^2/[2(1-\rho)], \text{ since } V(T) = 0 \\ &= \lambda t_0 + \lambda^2 t_0^2/[2(1-\lambda t_0)]. \\ &= (2\lambda t_0 - \lambda^2 t_0^2)/[2(1-\lambda t_0)]. \end{aligned}$$

This is the mean number in the queue left by a departing customer.

3.11 EXPECTED WAITING TIMES

The device of only looking at points of service termination will not help with the problems of expected waiting time. For this relates to the time from the moment of arrival to the moment of departure and we cannot assume that one customer arrives when another customer departs. In general, he will arrive when the server is busy and, if the service time is not exponential, the residual time to finish depends on its current duration. The subsequent difficulties can often be avoided by applying a general theorem due to Little [2]. This says that in an equilibrium queueing process, if $1/\lambda$ is the expected time between arrivals, $E(N)$ the mean number present and $E(W)$ the mean time spent by a customer from arrival to departure, then, under quite general conditions,

$$E(N) = \lambda E(W). \tag{3.28}$$

Some intuitive support for (3.28) in the case of Poisson arrivals at rate λ can be obtained by the following consideration. Consider a customer who spends time w in the system. Then the expected number arriving behind him, given w, in that time is λw. Hence the expected number arriving behind him is $\lambda E(W)$. But this is the expected number left by a departing customer *and if taken to be the expected number in the queue* is also $E(N)$. For example, if the service times are constant and have duration t_0 and customers arrive in a Poisson stream at rate λ, we have calculated the expected number left by a departing customer as

$$\lambda t_0 (2 - \lambda t_0)/[2(1 - \lambda t_0)].$$

Hence, after dividing by λ, Little's theorem gives the expected waiting time as

$$t_0 (2 - \lambda t_0)/[2(1 - \lambda t_0)]$$
$$= t_0 (2 - \rho)/[2(1 - \rho)] \text{ if } \rho = \lambda t_0 < 1.$$

Problem 15. For constant service time t_0 in the queue $M/D/1$, we might be prepared to accept that the average residual time for a customer found to be in service is $t_0/2$. If an arriving customer finds $n(> 0)$ in front, he must wait to commence service an average time $(n-1)t_0 + t_0/2$. Hence show that the expected queueing time is $\rho t_0/[2(1-\rho)]$. (Hint: if $n=0$ no residual service time.) ■

If Little's theorem is applied to (3.27), we obtain the Pollaczek–Khintchine formula

$$E(W) = \frac{1}{\lambda} E(N) = \frac{1}{\lambda} \left[\rho + \frac{\lambda^2 V(T) + \rho^2}{2(1-\rho)} \right]. \tag{3.29}$$

3.12 FURTHER DISCUSSION OF LITTLE'S THEOREM

The scope of this theorem is clearly very wide since, for instance, nothing is said about the number of servers. In the illustrative examples the service distributions are assumed to be independent and exponentially distributed with a common parameter μ and an equilibrium distribution is assumed.

Example 5. M/M/∞.
This system is a possible model for a self-service store. An arriving customer starts service immediately and his waiting time has expectation $1/\mu$. The system is equivalent to the death–immigration process discussed in Chapter 2, where we showed that the limiting number present had a Poisson distribution with mean (λ/μ), which satisfies $E(N) = \lambda E(W)$.

Problem 16. Verify Little's theorem for the queueing systems $M/M/1$ and $M/M/2$. ■

Example 6. M/M/s. Blocked customers lost
This system has been considered as a model for a telephone exchange which can take up to s calls. A caller who finds all s lines busy is blocked and his call is lost. Erlang showed that, in equilibrium, the probability, p_n, that just n lines are engaged is proportional to $(\lambda/\mu)^n/n!$ $(n = 0,1,\ldots,s)$. It is readily shown that the expected number of engaged lines is $(\lambda/\mu)(1-p_s)$. Now with probability p_s, all lines are engaged, and the caller gives up and has zero waiting time; while, with probability $1-p_s$, he makes connection and he holds a line for an average time $1/\mu$. Hence the expected waiting time is $(1-p_s)/\mu$ and Little's theorem is satisfied.

Problem 17. $M/M/1$. *With customer discouragement.* On finding n customers in front, a new arrival queues for service with probability $1/(n+1)$. Otherwise he leaves at once, and hence has zero waiting time. Show that Little's theorem holds. ■

Problem 18. $M/M/1$. *With finite capacity m.* The arriving customer only joins the queue if there are fewer than m in front. Show that Little's theorem is satisfied. ■

Example 6 and Problem 17 assume that the probability that an arriving customer finds the system full is equal to the outside observer's probability that the system is full. This is justified for Poisson arrivals (see section 3.5).

3.13 THE SYSTEM M/D/s

The mathematical difficulties associated with general service distribution and several servers would take us beyond the level of the present discussion. In the case of a single server, we know that if the next customer was already in line then he would start service as soon as the previous customer had departed. If there are several servers, this no longer applies since he has access to other servers. We have already discussed the case of exponential service times where this difficulty can be overcome (see Example 1).

In the case of constant service times t_0 some progress can be made by switching attention to the number queueing at times which are separated by an interval of t_0. That is we look at the queue at times t, $t + t_0$,.... We profit from observing that

(1)   all those receiving service at time t will have left by time $t+t_0$; and
(2)   all those arriving after t will still be there at time $t+t_0$, whether or not they received service immediately on arrival.

Let there be s servers, and let $p_j(t)$ be the probability that at time t there are j customers in the queue (waiting or being served). There will be $j > 0$ present at time $t + t_0$ if either

(a)   there were s or fewer present at t, all of these were served by $t + t_0$ and j arrived in time t_0; or
(b)   there were $s + k$ ($k \geqslant 1$) present at t, s of these were served by $t + t_0$ and $j - k$ arrived in time t_0.

Hence, if π_j is the probability of j arrivals in an interval t_0,

$$p_j(t+t_0) = \sum_{i=0}^{s} p_i(t)\,\pi_j + \sum_{k=1}^{j} p_{s+k}(t)\,\pi_{j-k}, \; j>0.$$

Suppose as $t \to \infty$ that an equilibrium distribution exists so that $p_j(t + t_0) \to p_j$, then

$$p_j = \sum_{i=0}^{s} p_i\pi_j + \sum_{k=1}^{j} p_{s+k}\,\pi_{j-k}, \; j>0. \tag{3.30}$$

Similarly for $j = 0$,

$$p_0 = \sum_{i=0}^{s} p_i\pi_0. \tag{3.31}$$

Let $G(\theta) = \sum_{j=0}^{\infty} p_j\theta^j$, $H(\theta) = \sum_{j=0}^{\infty} \pi_j\theta^j$. Multiply equation (3.30) by θ^j and sum over j from 1 to ∞ to obtain

$$\sum_{j=1}^{\infty} p_j\theta^j = \left(\sum_{i=0}^{s} p_i\right) \sum_{j=1}^{\infty} \pi_j\theta^j + \sum_{j=1}^{\infty} \sum_{k=1}^{j} p_{s+k}\,\theta^j\pi_{j-k},$$

that is

$$G(\theta) - p_0 = \left(\sum_{i=0}^{s} p_i\right)[H(\theta) - \pi_0] + \sum_{k=1}^{\infty} p_{s+k}\,\theta^k \sum_{j=k}^{\infty} \theta^{j-k}\pi_{j-k},$$

$$G(\theta) - p_0 = \left(\sum_{i=0}^{s} p_i\right) [H(\theta) - \pi_0] + \theta^{-s} \sum_{k=1}^{\infty} p_{s+k} \theta^{s+k} . H(\theta). \tag{3.32}$$

In virtue of (3.31),

$$G(\theta) = \left(\sum_{i=0}^{s} p_i\right) H(\theta) + \theta^{-s} . H(\theta) \, [G(\theta) - \sum_{i=0}^{s} p_i \theta^i]. \tag{3.33}$$

In principle then, (3.33) is available for computing various measures for the system.

Problem 19. Show that if $s = 1$, (3.33) reduces to (3.21). ∎

This section really has in mind that the customers arrive in a Poisson stream at rate λ. Hence, for the interval t_0,

$$\pi_r = \frac{(\lambda t_0)^r \exp(-\lambda t_0)}{r!}$$

$$H(\theta) = \sum_{r=0}^{\infty} \frac{(\lambda \theta t_0)^r \exp(-\lambda t_0)}{r!}$$

$$= \exp[-\lambda t_0 (1-\theta)] = \exp[-\rho(1-\theta)]$$

if $\rho = \lambda t_0$.

We now have to find $p_i (i = 0,1,2,\ldots,s-1)$. Note that the term in p_s cancels in (3.33). Thus, if $s = 2$,

$$G(\theta) = (p_0 + p_1) H(\theta) + \theta^{-2} H(\theta) \, [G(\theta) - p_0 - p_1 \theta],$$

and, solving for $G(\theta)$,

$$G(\theta) \, [\theta^2 - H(\theta)] = H(\theta) \, [p_0(\theta^2 - 1) + p_1 \theta(\theta-1)],$$

$$G(\theta) = \frac{H(\theta) \, [p_0(\theta+1) + p_1 \theta] \, (\theta-1)}{\theta^2 - H(\theta)}. \tag{3.34}$$

Now when $H(\theta) = \exp[-\rho(1-\theta)]$, $H(\theta)$ is increasing with θ and hence $\theta^2 - H(\theta)$ has just two zeros. Clearly $\theta = 1$ is one of these — let $\theta = \theta_0$ be the other. But for $|\theta| \leqslant 1$, $G(\theta)$ is bounded. Hence the numerator of (3.34) must be zero at $\theta = 1$ and $\theta = \theta_0$. Hence $p_0(\theta_0 + 1) + p\theta_0 = 0$. The limit as $\theta \to 1$ of (3.34) is $G(1) = 1$ and provides another equation for p_0, p_1.

Problem 20. If $H(\theta) = \exp[-\rho(1-\theta)]$, $\rho = \frac{1}{2}$, verify that approximately, $p_0 = 0.60, p_1 = 0.30$. ∎

*3.14 GENERAL ARRIVALS AND EXPONENTIAL SERVERS ($GI/M/s$)

Suppose there are s servers, with independent service times each distributed exponentially with parameter μ, and the p.d.f. for inter-arrival times is $f(t)$. Instead of looking at the system at moments of completed service, we now consider the situation at moments of arrival of a customer. In this way, we can exploit the fact that however many servers are busy, the residual service times are again exponential with the same parameter μ. Let then p_{ij} be the conditional probability that an arriving customer finds j before him (waiting or being served), given that the previous customer found i. Now that same arriving customer cannot find more than $i + 1$ in front. Hence $p_{ij} = 0, j > i + 1$. There are three cases to consider, according to whether either customer in question did or did not find all the servers busy.

(a) $j \leqslant i + 1 \leqslant s$. Neither customer found anybody waiting for service though not all the servers were necessarily busy. In that case, $i + 1 - j$ must have finished in the course of one inter-arrival time. If the inter-arrival time is t, the probability that any particular service lasts longer than t is $\exp(-\mu t)$. Hence we have $(i + 1)$ binomial trials with parameter $\exp(-\mu t)$ and the probability of just j 'successes' is

$$\binom{i+1}{j} e^{-\mu t j} (1 - e^{-\mu t})^{i+1-j}.$$

But this probability must be integrated over the distribution of the arrival time. Hence

$$p_{ij} = \int_0^\infty \binom{i+1}{j} \exp(-\mu t j)\,(1 - \exp(-\mu t))^{i+1-j} f(t)\mathrm{d}t. \tag{3.35}$$

Note that if $s = 1, i + 1 \leqslant s = 1$ only permits $i = 0$. Thus $j \leqslant i + 1$ allows $j = 0,1$.

(b)   If all the servers are busy throughout the inter-arrival time, $s \leqslant j \leqslant i + 1$. The system now behaves as though there were a single service distribution but with parameter $s\mu$. Hence the probability of $(i + 1 - j)$ services completed in time t is

$$\frac{(s\mu t)^{i+1-j}}{(i+1-j)!} \exp(-s\mu t).$$

Hence, allowing for the distribution of the inter-arrival time,

$$p_{ij} = \int_0^\infty \frac{(s\mu t)^{i+1-j}}{(i+1-j)!} \exp(-s\mu t) f(t)\mathrm{d}t. \tag{3.36}$$

Notice that $i = s - 1, j = s$ is a 'boundary' case for both (3.35), (3.36).

(c)   If $j < s < i + 1$, then all the servers are busy for part of the inter-arrival time, certainly until $i + 1 - s$ have completed service. Since the time for some customer to leave has an exponential distribution with parameter μs, the time for $i + 1 - s$ to leave has the gamma distribution $\Gamma(i+1-s, \mu s)$. Suppose the time until the $(i + 1 - s)$th departure is v. The remaining $(i + 1 - j) - (i + 1 - s) = s - j$ departures obey the binomial pattern as in (a) in time $t - v$ if the inter-arrival time is t. Since v, t must be integrated out, subject to $v \leqslant t$, p_{ij} is

$$\int_0^\infty \int_0^t \frac{s\mu(s\mu v)^{i-s}}{(i-s)!} e^{-s\mu v} \binom{s}{j} e^{-j\mu(t-v)} [1-e^{-\mu(t-v)}]^{s-j} f(t) dv dt. \qquad (3.37)$$

As an example of this somewhat complicated formula, consider $s = 1$. Then $i + 1 >$ is satisfied for $i > 0$ but $j < 1$ only for $j = 0$. If the inter-arrival time is constant $= t_0$, as in an appointment system, $0 \leqslant v \leqslant t_0$ and

$$p_{i0} = \int_0^{t_0} \frac{\mu(\mu v)^{i-1}}{(i-1)!} e^{-\mu v} [1 - e^{-\mu(t_0-v)}] dv, \quad i \geqslant 1. \qquad (3.38)$$

Problem 21. Show that, in (3.38),

$$p_{i0} = \sum_{r=i+1}^\infty \frac{(\mu t_0)^r}{r!} e^{-\mu t_0}. \quad \blacksquare$$

*3.15 EQUILIBRIUM DISTRIBUTION

If $p_j^{(m)}$ is the probability that customer number m finds j in front on arrival, then

$$p_j^{(m)} = \sum_{i=j-1}^\infty p_i^{(m-1)} p_{ij}, \quad j \geqslant 1,$$

$$p_0^{(m)} = \sum_{i=0}^\infty p_i^{(m-1)} p_{i0}.$$

If $p_j^{(m)} \to p_j$ as $m \to \infty$, we have the equilibrium equations

$$p_j = \sum_{i=j-1}^\infty p_i p_{ij}, \quad j \geqslant 1, \qquad (3.39)$$

$$p_0 = \sum_{i=0}^\infty p_i p_{10}. \qquad (3.40)$$

D. G. Kendall [3] has shown that (3.39) has a solution which, from p_{s-1} onwards, is essentially geometric — that is

$$p_j = c\alpha^j, j \geqslant s - 1, \text{ where } 0 < \alpha < 1.$$

There are thus two constants to find, c and α. To identify α, consider the equation for p_s.

$$p_s = \sum_{i=s-1}^{\infty} p_i p_{is}$$

that is

$$c\alpha^s = \sum_{i=s-1}^{\infty} c\alpha^i p_{is}. \tag{3.41}$$

But here we have $i \geqslant s - 1$ and $j = s$ and hence p_{is} is supplied by (3.36). After substituting in (3.41)

$$c\alpha^s = \sum_{i=s-1}^{\infty} c\alpha^i \int_0^{\infty} \frac{(\mu s t)^{i+1-s}}{(i+1-s)!} \exp(-\mu s t) f(t) dt. \tag{3.42}$$

(3.42) may be written

$$\alpha = \int_0^{\infty} \sum_{i=s-1}^{\infty} \frac{(\alpha \mu s t)^{i+1-s}}{(i+1-s)!} \exp(-\mu s t) f(t) dt,$$

so that

$$\alpha = \int_0^{\infty} \exp(\alpha \mu s t - \mu s t) f(t) dt. \tag{3.43}$$

Example 7
Suppose that we have exponential inter-arrival times with parameter λ, i.e. the system $M/M/s$. Here $f(t) = \lambda \exp(-\lambda t)$.

$$\int_0^{\infty} \lambda \exp[-t(\mu s - \mu s \alpha + \lambda)] dt$$

$$= \frac{\lambda}{\mu s - \mu s \alpha + \lambda}.$$

But this is required to be α, hence

$$\alpha = \lambda/(\mu s - \mu s \alpha + \lambda)$$

or

$$\mu s \alpha - \mu s \alpha^2 + \alpha \lambda = \lambda.$$

$$\mu s \alpha(\alpha - 1) - \lambda(\alpha - 1) = 0.$$

We seek the root $\neq 1$, that is $\alpha = \lambda/\mu s$. For an equilibrium distribution to exist $\rho = \lambda/\mu s < 1$, hence ρ is the required root.

Problem 22. (3.43) is of the form $\alpha = \phi(\alpha)$. Verify that $\phi(0) > 0$, $\phi(1) = 1$, and that $\phi'(1) > 1$ provided that $1/\rho = \mu s E(T) > 1$ where T is the inter-arrival time between customers. Draw a sketch of the graphs of $\beta = \alpha$, $\beta = \phi(\alpha)$ and confirm that when $\rho < 1$, there is another value α such that $\alpha = \phi(\alpha)$ and $\alpha \neq 1$. ∎

The rest of the calculation is somewhat tiresome, for we still have to find $p_0, p_1, \ldots, p_{s-2}$ and c. This is done in principle by starting with the equation for p_{s-1}

$$p_{s-1} = \sum_{i=s-2}^{\infty} p_i p_{i,s-1}. \tag{3.44}$$

In (3.44), we have $p_i = c\alpha^i$, $i \geqslant s-1$ and the $p_{i,s-1}$ can be calculated, hence we solve for p_{s-2} and so on.

Example 8

Let $s = 1$, and let the arrival distribution be exponential with parameter λ, or $f(t) = \lambda e^{-\lambda t}$. Actually there is little to do here since from the previous example $\alpha = \lambda/\mu$ and $p_i = c\alpha^i$, $i \geqslant 0$. Hence c can be found via $\sum_{i=0}^{\infty} p_i = 1$, $c = 1 - \alpha = 1 - \lambda/\mu$. However, it is of interest to see how the computations generally required are carried out. Since $s = 1$, p_{s-1} is p_0 and, from (3.40)

$$p_0 = \sum_{i=0}^{\infty} p_i p_{i0}.$$

In (3.37), $s = 1$, $j = 0$, $i \geqslant 1$, $f(t) = \lambda \exp(-\lambda t)$,

$$p_{i0} = \int_0^{\infty} \int_0^t \frac{\mu(\mu\nu)^{i-1}}{(i-1)!} e^{-\mu\nu} [1 - e^{-\mu(t-\nu)}] \lambda e^{-\lambda t} d\nu dt.$$

$$= \int_0^{\infty} \lambda \sum_{r=i+1}^{\infty} \frac{(\mu t)^r}{r!} e^{-(\mu+\lambda)t} dt, \text{ using the result of Problem 21,}$$

$$= \lambda \sum_{r=i+1}^{\infty} \frac{\mu^r}{r!} \int_0^{\infty} t^r e^{-(\mu+\lambda)t} dt$$

$$= \lambda \sum_{r=i+1}^{\infty} \frac{\mu^r}{r!} \left[\frac{r!}{(\lambda+\mu)^{r+1}} \right]$$

$$= \left(\frac{\mu}{\lambda+\mu} \right)^{i+1}.$$

Hence, from (3.40),

$$\sum_{i=0}^{\infty} p_i p_{i0} = p_0 p_{00} + \sum_{i=1}^{\infty} p_i p_{i0}$$

$$= p_0 p_{00} + \sum_{i=1}^{\infty} c\alpha^i \left(\frac{\mu}{\lambda+\mu}\right)^{i+1}$$

$$= p_0 p_{00} + \frac{c\mu}{\lambda+\mu} \cdot \frac{\alpha\mu/(\lambda+\mu)}{1-\alpha\mu/(\lambda+\mu)} .$$

That is, since $\alpha = \lambda/\mu$, and from (3.35) $p_{00} = \mu/(\lambda+\mu)$,

$$p_0 = p_0 \frac{\mu}{\lambda+\mu} + \frac{c\lambda}{\lambda+\mu},$$

or

$$p_0 = c.$$

Problem 23. For a single server with exponential service time with parameter μ, customers arrive at constant intervals t_0 and $\mu t_0 > 1$. If the limiting probability, p_j, of an arriving customer finding j customers in front is of the form $c\alpha^j (j \geq 1)$, use the relation $\alpha = \exp[-\mu t_0(1-\alpha)]$ to verify that $p_0 = c$.

{Hint: use $p_{i0} = \sum_{r=i+1}^{\infty} \frac{(\mu t_0)^r}{r!} \exp(-\mu t_0)$}. ∎

*3.16 QUEUEING TIME: *GI/M/s* QUEUE

In the equilibrium distribution, if we are only interest in the queueing time then we can manage without finding c in the relation $p_j = c\alpha^j$. An arriving customer will have to queue if there are at least s customers in front, with probability

$$\sum_{j=s}^{\infty} c\alpha^j = c\alpha^s/(1-\alpha).$$

Hence the conditional probability of finding j in front, given that he has to queue at all, is

$$\frac{c\alpha^j}{c\alpha^s/(1-\alpha)} = (1-\alpha)\alpha^{j-s}, j \geq s,$$

and does not depend on c. If he has to queue, given j in front, then $j - s + 1$ exponential services, each with paramter $s\mu$, must end before he can commence his own service. But the sum of these has a $\Gamma(j-s+1, s\mu)$ distribution. Hence the density of the queueing time, q, give that it is positive, is

$$\sum_{j=s}^{\infty} \frac{s\mu(s\mu q)^{j-s}\exp(-s\mu q)}{(j-s)!} \cdot (1-\alpha)\alpha^{j-s}$$

$$= s\mu(1-\alpha) \exp[-qs\mu(1-\alpha)] .$$

This is the p.d.f. of an exponential distribution with mean $1/[s\mu(1-\alpha)]$. The overal average queueing time requires multiplying this mean by the probability of having to queue at all, and is

$$\frac{c\alpha^s}{s\mu(1-\alpha)^2},$$

which does of course involve c. For example, if $s = 1$, then $c = 1-\alpha$ and the probability of queueing is α. The average queueing time is $\alpha/[(1-\alpha)\mu]$. Further, if customers arrive at constant intervals t_0 and $\mu t_0 = 2$, then $\alpha = \exp[-2(1-\alpha)]$ and from tables of the exponential distribution, $\alpha = 0.203$. Fortunately, tables are available for c for the more common inter-arrival distributions.

3.17 SERVICING MACHINES

We propose to give a limited account of the problems of machine maintenance and that under only the simplest assumptions. It is then seen to bear a strong resemblance to certain situations in queueing theory. Suppose then that m machines are maintained by r repairmen $(r < m)$. The running times for machines have independent exponential distributions with the same parameter λ and repair times have independent exponential distributions with a common parameter μ. If $p_n(t)$ is the probability that, at time t, n machines are not working, and bearing in mind that if $n < r$ all failed machines can receive attention, the usual forward equations are

$$\frac{dp_n(t)}{dt} = -[(m-n)\lambda+n\mu]p_n(t) + (m-n+1)\lambda p_{n-1}(t) + (n+1)\mu p_{n+1}(t). \qquad (3.45)$$

On the other hand, if $n > r$, some failed machines must wait for attention and we have,

$$\frac{dp_n(t)}{dt} = -[(m-n)\lambda+r\mu]p_n(t) + (m-n+1)\lambda p_{n-1}(t) + r\mu p_{n+1}(t). \qquad (3.46)$$

The boundary cases, $n = 0, r, m$ should be considered separately (see Problem 24).

Problem 24. Show that $p_0(t)$ satisfies (3.45), provided that $p_{-1}(t) \equiv 0$. Show further that $p_r(t), p_m(t)$ satisfy (3.46). ∎

Assuming that an equilibrium distribution exists in which $p_n(t) \to p_n$ as $t \to \infty$, then from (3.45), (3.46) we have

$$\lambda[(m-n)p_n - \{m-(n-1)\}p_{n-1}] = \mu[(n+1)p_{n+1} - np_n], 0 \leqslant n < r \qquad (3.47)$$

and

$$\lambda[(m-n)p_n - \{m-(n-1)\}p_{n-1}] = \mu r[p_{n+1} - p_n], \qquad r \leqslant n \leqslant m. \qquad (3.48)$$

We have displayed (3.47), (3.48) in a form well suited to summing over n. After taking care to switch equations at $n = r$, it is readily verified that

$$(m-n)\lambda p_n = (n+1)\mu p_{n+1}, \quad 0 \leqslant n < r; \tag{3.49}$$

$$(m-n)\lambda p_n = r\mu p_{n+1}, \quad r \leqslant n \leqslant m. \tag{3.50}$$

Among measures which have been proposed for evaluating such systems are

(1) coefficient of loss by machines = expected proportion of machines waiting for repair work; and
(2) coefficient of loss by repairmen = expected proportion of idle repairmen.

In computing these coefficients, it should be born in mind that:

$$E \text{ (number of failed machines waiting for attention)} = \sum_{n=r+1}^{m} (n-r)p_n,$$

$$E \text{ (number of idle repairmen)} = \sum_{n=0}^{r-1} (r-n)p_n.$$

Problem 25. If $m = 4$, $r = 2$, $\mu = 2\lambda$, show that the equilibrium probability that all the machines are working is 16/87. Calculate the expected number of idle operators and the expected number of machines awaiting repair. ∎

Problem 26. If $r = 1$, show that the expected number of machines not working after a long time is $m - \mu(1-p_0)/\lambda$. (Hint: consider summing the appropriate equation over n.) ∎

REFERENCES

[1] *Lecture Notes on Queueing Systems*, B. Conolly. Ellis Horwood Ltd., Chichester, 1975.
[2] 'A proof for the queueing formular: $L = \lambda W$', J. D. C. Little. *Operations Research*, **9** (3) (1961).
[3] Stochastic processes occurring in the theory of queues. D. G. Kendall. *Ann. Math. Stat.* (1953).

BRIEF SOLUTIONS AND COMMENTS ON THE PROBLEMS

Problem 1. Not as tiresome as it appears, if the hint is used.

$$\phi_{n+k}(t) = \exp[-(\lambda+\mu)t] \, (\lambda/\mu)^{(n+k)/2} \, I_{n+k} [2t(\lambda\mu)^{1/2}],$$

$$\phi'_{n+k}(t) = -(\lambda+\mu) \, \phi_{n+k}(t) + \exp[-(\lambda+\mu)t] \, (\lambda/\mu)^{(n+k)/2}$$
$$I'_{n+k} [2t(\lambda\mu)^{1/2}] \, 2(\lambda\mu)^{1/2}.$$

Now use $I_n'(y) = \frac{1}{2}[I_{n-1}(y) + I_{n+1}(y)]$ and show that

$$\phi_{n+k}'(t) = -(\lambda+\mu)\phi_{n+k}(t) + \lambda\phi_{n+k-1}(t) + \mu\phi_{n+k+1}(t).$$

Finally differentiate the suggested $p_{on}(t)$, and substitute for $\phi_{n+k}'(t)$, $\phi_{n+k+2}'(t)$.

Problem 2. In the interval $(0,\delta t)$ there is either:

(i) one arrival, with approximate probability $\lambda\delta t$, there are then two in the system and n after a further interval $t-\delta t$,

(ii) one service terminates, with probability approximately $\mu\delta t$, there is no-one present but n after a further interval $t-\delta t$,

(iii) neither event takes place, one person is still present and then n after a further interval $t-\delta t$.

$p_{1n}(t) = p_{2n}(t-\delta t)\ \lambda\delta t + p_{on}(t-\delta t)\ \mu\delta t + p_{1n}(t-\delta t)\ (1-\lambda\delta t)\ (1-\mu\delta t)$. For $p_{on}(t)$, a service cannot terminate in $(0,\delta t)$.

Problem 3. Multiplying the first equation by θ^n and summing from 1 to ∞ we write:

$$\rho \sum_1^\infty p_{n-1}(t)\theta^n = \rho\theta \sum_1^\infty p_{n-1}(t)\theta^{n-1} = \rho\theta G;$$

$$(1+\rho) \sum_1^\infty p_n(t)\theta^n = (1+\rho)\ [G-p_0(t)]\ ;\ \text{and}$$

$$\sum_1^\infty p_{n+1}(t)\theta^n = \frac{1}{\theta} \sum_1^\infty p_{n+1}(t)\theta^{n+1} = [G-p_0(t) - \theta p_1(t)]/\theta.$$

The second equation helps eliminate some terms. The hint assumes that differentiation with respect to t and with respect to θ can be interchanged.

Problem 4. By considering the intervals $(0,t)$, $(t,t+\delta t)$

$$p_{10}(t+\delta t) = p_{10}(t)\ (1-\lambda\delta t) + p_{11}(t)\mu\delta t + o(\delta t).$$

After using the indicated integrating factor, use the initial condition $p_{10}(0) = 0$.

Problem 5. Since T_n-T_1 has the $\Gamma(n-1,\lambda)$ distribution and T_1 is $\Gamma(1,\lambda)$, the joint p.d.f. of T_1,T_n is

$$\lambda \exp(-\lambda t_1) \cdot \lambda\ [\lambda(t_n-t_1)]^{n-2}\ \exp[-\lambda(t_n-t_1)]/(n-2)!\ .$$

The distribution of T_n is $\Gamma(n,\lambda)$ and has p.d.f.

$$\lambda(\lambda t_n)^{n-1}\ \exp(-\lambda t_n)/(n-1)!\ .$$

The conditional p.d.f. of T_1 given T_n is found by dividing the joint p.d.f. of T_1, T_n by the marginal p.d.f. of T_n.

On the other hand, the number, $N(t_n - t_1)$, of customers arriving in $t_n - t_1$ has a Poisson distribution with parameter $\lambda(t_n - t_1)$. Hence the (mixed) p.d.f. of T_1 and $Pr[N(t_n - t_1) = n - 2]$ is

$$\lambda \exp(-\lambda t_1) \, [\lambda(t_n - t_1)]^{n-2} \exp[-\lambda(t_n - t_1)]/(n-2)! \, .$$

The distribution of $N(t_n)$ is Poisson, with parameter λt_n. Hence $Pr[N(t_n) = n-1] = (\lambda t_n)^{n-1} \exp(-\lambda t_n)/(n-1)!$. After division, the required result follows.

Problem 6. If he finds n in front then the queueing time has a $\Gamma(n,\mu)$ distribution. In the equilibrium distribution, the probability of n is $(1-\rho)\rho^n$. Hence the queueing time for $n > 0$ has p.d.f.

$$\sum_{n=1}^{\infty} \frac{\mu(\mu t)^{n-1} \exp(-\mu t)\rho^n \, (1-\rho)}{(n-1)!}$$

$$= \mu(1-\rho)\rho \exp[-\mu(1-\rho)t] \, .$$

This is *not* the p.d.f. of an exponential distribution. But since the probability of zero queueing time is $1-\rho$, the *conditional* p.d.f., given that it is positive, is found by dividing by ρ.

Problem 7

$$Pr(N \geqslant s) = \frac{s^s}{s!} \sum_{n=s}^{\infty} \left(\frac{\lambda}{s\mu}\right)^n p_0 = \left(\frac{\lambda}{\mu}\right)^s \frac{p_0}{s!} / (1 - \lambda/s\mu).$$

$$Pr(N = n | N \geqslant s) = Pr(N = n \text{ and } n \geqslant s)/Pr(N \geqslant s).$$

Problem 8. From the results of Example 1 we have for $s = 2$,

$$p_n = 2\left(\frac{\lambda}{2\mu}\right)^n p_0, \quad n \geqslant 2;$$

$$p_n = \frac{1}{n!}\left(\frac{\lambda}{\mu}\right)^n p_0, \quad n = 0,1, \text{ that is } p_1 = \left(\frac{\lambda}{\mu}\right) p_0.$$

Since the probabilities must total 1,

$$1 = p_0 + \frac{\lambda}{\mu} p_0 + 2 \sum_{2}^{\infty} \left(\frac{\lambda}{2\mu}\right)^n p_0 = p_0 \left[1 + \frac{\lambda}{\mu} + \frac{2(\lambda^2/4\mu^2)}{1 - \lambda/2\mu}\right].$$

Hence $p_0 = (2\mu - \lambda)/(2\mu + \lambda)$. For the expected number, observe that $\sum_{1}^{\infty} n\rho^{n-1} = (1-\rho)^{-2}$.

Problem 9. The second summation in (3.11) is $\left(\dfrac{\lambda/\mu}{s!}\right)^s \sum\limits_{n=s}^{\infty} \left(\dfrac{\lambda}{s\mu}\right)^{n-s} =$

$\left(\dfrac{\lambda/\mu}{s!}\right)^s \cdot \dfrac{1}{1-(\lambda/s\mu)} \cdot$ But $\left(\dfrac{\lambda/\mu}{s!}\right)^s \to 0$ as $s \to \infty$. Hence

$$p_0 \to \left[\sum_0^{\infty} (\lambda/\mu)^n/n!\right]^{-1} = \exp(-\lambda/\mu).$$

Problem 10. From (3.9), dropping m, $p_n = \prod\limits_{r=1}^{n} (\lambda_{r-1}/\mu_r)p_0 = \lambda^n p_0/(n!\mu^n)$.

Since $\sum\limits_{1}^{\infty} p_n = 1-p_0$, $p_0 [\exp(\lambda/\mu)-1] = 1-p_0$. Hence $p_0 = \exp(-\lambda/\mu)$. The reader should confirm that the model $\lambda_n = \lambda$, $\mu_n = n\mu$ leads to the same result.

Problem 11

$$p_n = \prod\limits_{r=1}^{n} (\lambda_{r-1}/\mu_r)p_0 = (\lambda/\mu)^n p_0, \qquad 1 \leqslant n \leqslant m. \text{ But } p_n = 0 \text{ for } n > m$$

$$1 = p_0 \sum\limits_{0}^{m} (\lambda/\mu)^n = p_0 (1-\rho^{m-1})/(1-\rho), \text{ where } \rho = \lambda/\mu.$$

Problem 12. If r are active, then $s-r$ are idle ($r \leqslant s$). Since in the equilibrium distribution, each server has the same probability of being idle, the conditional probability that a particular server is idle is $(s-r)/(s)$. It is of course zero if the number in the system exceeds s. Hence the unconditional probability is $\sum\limits_{0}^{s-1}$ $(s-r)p_r/s = 1 - \rho$ from (3.14) where $\rho = \lambda/s\mu$.

Problem 13. After interchanging summation and integration,

$$H(\theta) = \int_0^{\infty} \left\{\mu \exp[-(\lambda+\mu)t] \sum\limits_{r=0}^{\infty} (\lambda t\theta)^r/r!\right\} dt$$

$$= \int_0^{\infty} \mu \exp[-(\lambda+\mu-\lambda\theta)t] \, dt$$

$$= \mu/(\lambda+\mu-\lambda\theta).$$

Notice that since $\lambda/\mu < 1$, $|\theta| \leqslant 1$, $\lambda+\mu-\lambda\theta > 0$, ensuring proper behaviour at the upper terminal of the integral.

Problem 14. Interchanging summation and integration

$$H(\theta) = \int_0^\infty \left\{ \mu^2 t \exp[-(\lambda+\mu)t] \sum_{r=0}^\infty (\lambda t\theta)^r/r! \right\} dt$$

$$= \int_0^\infty \mu^2 t \exp[-(\lambda+\mu-\lambda\theta)t] \, dt$$

$$= \{\mu/[\mu+\lambda(1-\theta)]\}^2.$$

That this is the square of the generating function in Problem 13 is no accident. The service distribution is the sum of two exponential distributions each with parameter μ. Now, from (3.21)

$$G(\theta) = p_0 H(\theta) (\theta-1)/[\theta-H(\theta)]$$

in which we substitute $H(\theta)$ and after a little elementary simplification, contrive to cancel the factor $(\theta-1)$. This leaves us with the required form.

$$G(1) = 1 = p_0 \mu^2/(\mu^2-2\lambda\mu), \text{ or } p_0 = 1 - 2\lambda/\mu.$$

The mean number of customers left is

$$\left[\frac{dG(\theta)}{d\theta} \right]_{\theta=1} = \lambda(2\mu-\lambda)/[\mu(\mu-2\lambda)].$$

Problem 15. Since $(n-1)t_0 + (t_0/2) = (n-1/2)t_0$, this is the queueing time for $n > 0$. When $n = 0$, the queueing time is zero. Hence expected queueing time is

$$t_0 \sum_1^\infty (n-\tfrac{1}{2})p_n = t_0 \left[\sum_1^\infty np_n - \tfrac{1}{2}(1-p_0) \right].$$

From section 3.10, Example 4, $\sum_1^\infty np_n = (2\rho-\rho^2)/[2(1-\rho)]$, and $p_0 = 1-\rho$ where $\rho = \lambda t_0$. On substituting, the required result is obtained. The corresponding waiting time is found by adding on another t_0, yielding $(2-\rho)t_0/[2(1-\rho)]$.

Problem 16
(a) $M/M/1$. In section 3.4 we have $E(N) = \rho/(1-\rho) = \lambda/(\mu-\lambda)$. In section 3.5, we find $E(W) = 1/(\mu-\lambda)$. The result follows.
(b) $M/M/2$. From Problem 8, $E(N) = 2\rho/(1-\rho^2)$ and from section 3.7, $E(W) = 1/[\mu(1-\rho^2)]$. This gives the required result, noting that $\rho = \lambda/2\mu$.

Problem 17. Problem 10 derives the equilibrium distribution of N as Poisson with parameter λ/μ. Hence $E(N) = \lambda/\mu$. Given that an arriving customer remains and finds n present, he stays for an expected time of $(n+1)/\mu$ before finishing service. The probability that he remains is $1/(n+1)$. With probability $n/(n+1)$ he does not stay and spend zero time in the system. Hence his expected waiting time is

$$\frac{(n+1)}{\mu} \cdot \frac{1}{n+1} + 0 \cdot \frac{n}{n+1} = \frac{1}{\mu},$$

for every n.

Problem 18. The solution to Problem 11 gives $p_n = (\lambda/\mu)^n p_0$. We do not trouble to evaluate p_0. An arriving customer finds n present and, waits an expected time $(n+1)/\mu$ to finish service, with probability p_n ($n \leqslant m-1$). With probability p_m he leaves at once.

$$E(W) = \sum_0^{m-1} \frac{n+1}{\mu} \left(\frac{\lambda}{\mu}\right)^n p_0.$$

$$E(N) = \sum_1^m n \left(\frac{\lambda}{\mu}\right)^n p_0 = \sum_0^{m-1} (n+1)\left(\frac{\lambda}{\mu}\right)^{n+1} p_0 = \lambda E(W).$$

Problem 19. If $s = 1$,

$$G(\theta) = (p_0+p_1)H(\theta)+H(\theta) \ [G(\theta)-p_0-p_1\theta]/\theta,$$

$$G(\theta) \ [\theta-H(\theta)] = p_0(\theta)H(\theta) \ (\theta-1), \text{ which is (3.21).}$$

Problem 20. To evaluate $G(1)$, use L'Hôpital's rule.

$$G(1) = H(1) \ (2p_0+p_1)/[2-H'(1)],$$

that is

$$1 = (2p_0+p_1)/[2-H'(1)].$$

But $H(\theta) = \exp[-\frac{1}{2}(1-\theta)]$ and $H'(1) = \frac{1}{2}$.

Clearly $p_0 = 0.60$, $p_1 = 0.30$ satisfies $G(1) = 1$. Substituting these values in $p_0(\theta_0+1) + p_1\theta_0 = 0$, we have $\theta_0 = -2/3$. From tables of the exponential distribution, $\exp[-\frac{1}{2}(1+2/3)]$ is approximately 0.44, i.e. $H(-5/6)$ is approximately $(-2/3)^2$, so that θ_0 is an approximate root of $\theta^2 - H(\theta) = 0$.

Problem 21

$$\exp(-\mu v) \ \{1-\exp[-\mu(t_0-v)]\} = \exp(-\mu v)-\exp(-\mu t_0),$$

$$\int_0^{t_0} \frac{\mu(\mu v)^{i-1}}{(i-1)!} \exp(-\mu t_0)dv = \left(\frac{\mu t_0}{i!}\right)^i \exp(-\mu t_0), \text{ and}$$

$$\int_0^{t_0} \frac{\mu(\mu v)^{i-1}}{(i-1)!} \exp(-\mu v)dv = \left[\frac{(\mu v)^i}{i!} \exp(-\mu v)\right]_0^{t_0} + \int_0^{t_0} \frac{\mu(\mu v)^i}{i!} \exp(-\mu v)dv$$

$$= \frac{(\mu t_0)^i}{i!} \exp(-\mu t_0) + \int_0^{t_0} \frac{\mu(\mu v)^i}{i!} \exp(-\mu v)dv.$$

Repeated integration by parts of this last integral gives the result required, after collecting terms.

Problem 22

$$\phi(0) = \int_0^\infty \exp(-\mu t) f(t) dt > 0, \quad \phi(1) = \int_0^\infty f(t) dt = 1.$$

$$\phi'(1) = \int_0^\infty \mu s t f(t) dt = \mu s E(T) > 1 \text{ if } \rho < 1.$$

A sketch shows that $\phi(\alpha) > \alpha$ at the origin, $\phi(\alpha) = \alpha$ at $\alpha = 1$ and the slope of $\phi(\alpha)$ exceeds that of the line $\beta = \alpha$ at $(1,1)$.

Problem 23. From (3.40),

$$p_0 = \sum_{i=0}^\infty p_i p_{i0} = p_0 p_{00} + \sum_{i=1}^\infty p_i p_{i0}.$$

Hence

$$p_0 = p_0 p_{00} + \sum_{i=1}^\infty c\alpha^i p_{i0}.$$

$p_{00} = [1-\exp(-\mu t_0)]$. Using the hint, substitute for p_{i0}, interchange summation over r and i (noting that $r \geqslant i+1 \Rightarrow i \leqslant r-1$) we have,

$$\sum_{i=1}^\infty c\alpha^i p_{i0} = c \exp(-\mu t_0) \sum_{r=2}^\infty \frac{(\mu t_0)^r}{r!} \left[\sum_{i=1}^\infty \alpha^i \right],$$

$$= c \exp(-\mu t_0) \sum_2^\infty \frac{(\mu t_0)^r}{r!} \frac{(\alpha-\alpha^r)}{1-\alpha},$$

$$= \frac{c \exp(-\mu t_0)}{1-\alpha} \left[\alpha\exp(\mu t_0) - \exp(\mu\alpha t_0) + 1 - \alpha \right].$$

But from the given relation, $\alpha\exp(\mu t_0) = \exp(\mu\alpha t_0)$. The result follows.

Problem 24. All are working at time $t+\delta t$ if either this was the case at time t and none failed or just one had failed and this was repaired in the interval $(t,t+\delta t)$.

$$p_0(t+\delta t) = p_0(1-m\lambda\delta t) + p_1(t)\mu\delta t + o(\delta t).$$

All have failed at time $t+\delta t$ if either this was the case at time t and none have been repaired or one was working and this failed in the interval $(t,t+\delta t)$.

$p_m(t+\delta t) = p_m(t)(1-r\mu\delta t) + p_{m-1}(t)\lambda\delta t + o(\delta t)$. This fits (3.346) if $p_{m+1}(t) \equiv 0$. Similarly

$p_r(t+\delta t) = p_r(t)(1-r\mu\delta t)[1-(m-r)\lambda\delta t] + p_{r+1}(t)r\mu\delta t + p_{r-1}(t)(m-r+1)\lambda\delta t$.

Problem 25. $\mu = 2\lambda$, $m = 4$, $r = 2$. Use equation (3.49) for $n = 0,1$.

$n = 0$, $\quad 4\lambda p_0 = 2\lambda p_1$, $\quad p_1 = 2p_0$.

$n = 1$, $\quad 3\lambda p_1 = 2(2\lambda)p_2$, $p_2 = 3p_0/2$.

For $\quad n = 2,3$, use equation (3.50).

$n = 2$, $\quad 2\lambda p_2 = 2(2\lambda)p_3$, $p_3 = 3p_0/4$

$n = 3$, $\quad \lambda p_3 = 2(2\lambda)p_4$, $p_4 = 3p_0/16$.

Finally substitute in $p_0 + p_1 + p_2 + p_3 + p_4 = 1$.

Problem 26. The equations, for $r = 1$, are

$$m\lambda p_0 = \mu p_1,$$

$$(m-n)\lambda p_n = m\lambda p_n - \lambda n p_n = \mu p_{n+1}.$$

Sum the second equation from 1 to m.

$$m\lambda(1-p_0) - \lambda E(N) = \mu(1-p_0-p_1),$$

Now substitute $m\lambda p_0$ for μp_1.

CHAPTER 4

Renewal Theory

4.1 INTRODUCTION

Suppose we have an unlimited supply of new and similar light bulbs. One of these is selected, inserted in a socket and allowed to burn until it fails. It is then immediately replaced with another bulb – and so on. We extract certain features of this situation.

(i)   The 'life' of each bulb is a non-negative random variable.
(ii) The 'life' of each bulb has the same distribution.
(iii) The 'lives' of different bulbs are independent.

The underlying structure is that of a series of events (failures) ordered in time. The waiting time from any event to the next is a random variable with a common distribution and is independent of previous events. We now give a formal definition of such a process. Let $X_1, X_2, \ldots$ be a sequence of non-negative, mutually independent random variables with a common distribution. Let $T_n = \sum_{i=1}^{n} X_i$, then the sequence $T_1, T_2, \ldots$ is said to constitute an **ordinary renewal process**. The individual T_i are the renewal points and the number of renewal points in the interval $(0,t)$ is denoted by N_t. The X_i are also known as inter-renewal times. Since each X_i is non-negative, $E(X_i) \geqslant 0$. We shall restrict our discussion to those processes for which the X_i have continuous distributions and finite expectations. In some applications, it helps to think of X_i as the 'life' of a component and N_t as the number of components replaced.

Example 1
In an ordinary renewal process, suppose the distribution of each X_i is exponential with parameter λ. Then T_n, being the sum of n independent exponentially distributed variables has the $\Gamma(n,\lambda)$ distribution (see section 1.2). Hence the expected time to the nth renewal is n/λ. Moreover, the number of renewals in $(0,t)$ has a Poisson distribution with mean λt. Since the exponential distribution has no memory, the number of renewals in any interval of duration t has a Poisson distribution with mean λt. These results, for this, the most simple of renewal processes, arise from results already established in Chapter 1.

4.2 DISTRIBUTION OF THE NUMBER OF RENEWALS

It will be apparent that the distribution of the number of renewals is of some practical interest. If the functioning of an item of equipment, and its replacement on failure, can be modelled as an ordinary renewal process then the distribution of the number of renewals gives some guidance as to the number of spares which should be on hand for a given period. To compute the distribution when the inter-renewal times do not necessarily have an exponential distribution we commence with the simple observation that there are *at least n* renewals before t when, and only when, the time to the nth renewal occurs before t. That is to say,

$$N_t \geqslant n \leftrightarrow T_n \leqslant t. \tag{4.1}$$

Thus these events must have the same probability, that is

$$Pr[N_t \geqslant n] = Pr[T_n \leqslant t]. \tag{4.2}$$

Now if each inter-renewal time has cumulative distribution function (c.d.f.) $F(.)$, then

$$Pr(T_n \leqslant t) = F_n(t), \quad i = 1,2,\ldots$$

where $F_n(.)$ is the c.d.f. of the sum of n identically and independently distributed random variables. Note that $F_1(.)$ is here just $F(.)$. We should perhaps remark that we choose not to count the placing of the first component as a renewal. By another application of equation (4.2) we have

$$
\begin{aligned}
Pr(N_t = n) &= Pr(N_t \geqslant n) - Pr(N_t \geqslant n+1) \\
&= Pr(T_n \leqslant t) - Pr(T_{n+1} \leqslant t) \\
&= F_n(t) - F_{n+1}(t), \quad n \geqslant 1, \tag{4.3}
\end{aligned}
$$

and

$$Pr(N_t = 0) = 1 - F_1(t).$$

We are immediately in a position to confirm part of Example 1. For if T_n has the $\Gamma(n, \lambda)$ distribution,

$$F_{n+1}(t) = \int_0^t \frac{\lambda(\lambda x)^n \exp(-\lambda x)}{n!} \, dx.$$

Integrating by parts,

$$F_{n+1}(t) = -\frac{(\lambda x)^n \exp(-\lambda x)}{n!} \Bigg]_0^t + \int_0^t \frac{\lambda(\lambda x)^{n-1} \exp(-\lambda x)}{(n-1)!} \, dx$$

$$= -\frac{(\lambda t)^n \exp(-\lambda t)}{n!} + F_n(t),$$

so that $Pr(N_t = n) = (\lambda t)^n \exp(-\lambda t)/n!$, which is the probability density function of a Poisson distribution with parameter λt.

Problem 1. In an ordinary renewal process, X_i has the $\Gamma(2,\lambda)$ distribution with p.d.f. $f(x) = \lambda(\lambda x) \exp(-\lambda x)$. Show that

$$Pr(N_t = n) = \left[\frac{(\lambda t)^{2n}}{2n!} + \frac{(\lambda t)^{2n+1}}{(2n+1)!}\right] \exp(-\lambda t). \quad \blacksquare$$

4.3 EXPECTED NUMBER OF RENEWALS

We shall be particularly interested in the expected number of renewals in $(0,t)$ since this will cast some light on the average cost of replacing components. Now

$$F_n(t) = \sum_{i=1}^{\infty} i Pr(N_t = i)$$

$$= \sum_{i=1}^{\infty} \sum_{n=1}^{i} 1 . Pr(N_t = i), \quad \text{interchanging the order of summation,}$$

$$= \sum_{n=1}^{\infty} \sum_{i=n}^{\infty} Pr(N_t = i)$$

$$= \sum_{n=1}^{\infty} Pr(N_t \geqslant n)$$

$$= \sum_{n=1}^{\infty} F_n(t), \text{ from equation (4.2).} \tag{4.4}$$

Equation (4.4) appears to provide a reasonable method for computing $E(N_t)$, though the reader may be experiencing some misgivings as to the behaviour of the series $\sum_{1}^{\infty} F_n(t)$. We leave aside questions of convergence for the time being.

Example 2

If X_i has the $\Gamma(2,\lambda)$ distribution then $T_n = \sum_{1}^{n} X_i$ has the $\Gamma(2n,\lambda)$ distribution. Hence

$$F_n(t) = \int_0^t \frac{\lambda(\lambda x)^{2n-1} \exp(-\lambda x)}{(2n-1)!} dx$$

$$E(N_t) = \sum_{1}^{\infty} F_n(t)$$

$$= \sum_{1}^{\infty} \int_{0}^{t} \frac{\lambda(\lambda x)^{2n-1} \exp(-\lambda x)}{(2n-1)!} \, dx$$

$$= \lambda \int_{0}^{t} \left[\sum_{1}^{\infty} \frac{(\lambda x)^{2n-1}}{(2n-1)!} \right] \exp(-\lambda x) dx$$

$$= \lambda \int_{0}^{t} \left[\frac{\exp(\lambda x) - \exp(-\lambda x)}{2} \right] \exp(-\lambda x) dx$$

$$= \frac{\lambda}{2} \int_{0}^{t} [1 - \exp(-2\lambda x)] \, dx$$

$$= \frac{\lambda t}{2} - \frac{1}{4} [1 - \exp(-2\lambda t)].$$

Of the result in Example 2 we observe that the expected number of renewals is *not* just proportional to t. There is only one ordinary renewal process for which the expected number of renewals in $(0,t)$ is proportional to t for all t and that is when the inter-renewal times have common exponential distributions (see Example 1). Nevertheless, in Example 2, the limit as t increases of $E(N_t)/t$ is $\lambda/2$, where $2/\lambda$ is the mean of a $\Gamma(2,\lambda)$ distribution. This accords with the intuitive idea, that over a long period, the expected number of renewals should approach t/μ where μ is the mean of an individual inter-renewal time.

If we consider two adjacent intervals $(0,t_1)$, (t_1,t_2), then of course it remains true that the sum of the expected number of renewals in these intervals equals the expected number in the single interval $(0,t_2)$. Hence we can find the expected number of renewals as the difference $E(N_{t_2}) - E(N_{t_1}) = \Sigma F_n(t_2) - \Sigma F_n(t_1)$. This difference is *not* the same as $\Sigma F_n(t_2 - t_1)$. This is because the process is not ordinary for the interval (t_1,t_2) since at t_1 the component in use is not new. In terms of costs it may pay to interrupt the process at t_1 and install a new component, regardless of its age. This is the subject of Problem 2.

Problem 2. In the ordinary renewal process of Example 2 we determined $E(N_t)$ as $\{2\lambda t - [1 - \exp(-2\lambda t)]\}/4$. The cost of installing the initial component is c_0 and for subsequently replacing a failed component is c_1. Calculate the expected cost for components up to time t_0 and show that it will be cheaper, on average, to install a new component at time $t_0/2$, at cost c_0, if

$$4c_0 < c_1 [1 - \exp(-\lambda t_0)]^2. \quad \blacksquare$$

*4.4 LIMITING BEHAVIOUR OF $F_n(t)$.

We now consider the convergence of the series $\sum_{1}^{\infty} F_n(t)$. It is reasonable to assume that $F(t) < 1$ for finite t for this corresponds to there being no finite time by which a part is bound to have failed. Now we cannot have $\sum_{1}^{n} X_i = T_n \leqslant t$ if the largest X_i already exceeds t. On the other hand if the largest X_i is less than t we cannot conclude that $T_n < t$. Hence,

$$Pr(T_n \leqslant t) \leqslant Pr\,[\max(X_1, X_2, \ldots, X_n) \leqslant t]$$

$$= Pr(X_1 \leqslant t, X_2 \leqslant t, \ldots, X_n \leqslant t)$$

$$= \prod_{1}^{n} Pr(X_i \leqslant t).$$

$$= [F(t)]^n. \qquad (4.5)$$

From equation (4.2)

$$\lim_{n \to \infty} [Pr(N_t \geqslant n)] = \lim_{n \to \infty} [Pr(T_n \leqslant t)]$$

$$\leqslant \lim_{n \to \infty} [F(t)]^n$$

$$= 0, \text{ since } F(t) < 1.$$

We conclude that, while not impossible, there is no positive probability of an infinite number of renewals in $(0,t)$ for finite t. Moreover, from equation (4.5), $Pr(T_n \leqslant t) = F_n(t) \leqslant [F(t)]^n$ and for $F(t) < 1$, the geometric series $\sum_{1}^{\infty} [F(t)]^n$ converges. Therefore $\sum_{1}^{\infty} F_n(t)$ converges and is less than or equal to $\sum_{1}^{\infty} [F(t)]^n$ $= F(t)/[1-F(t)]$. In any case $\sum_{1}^{\infty} F_n(t)$ exceeds the first term. In summary, $E(N_t)$ is finite and satisfies

$$F(t) \leqslant E(N_t) \leqslant F(t)/[1-F(t)]. \qquad (4.6)$$

Problem 3. Assuming $F(t) < 1$, use equation (4.3) to show that

$$\sum_{1}^{\infty} n\,Pr(N_t = n) = \sum_{1}^{\infty} F_n(t).$$

The reader is reminded that if $F(t) < 1$, not only $[F(t)]^n \to 0$ as $n \to \infty$, but so also does $n[F(t)]^n$. ■

We have omitted the irritating exceptional case when $F(t) = 1$, say for $t > t_0$. This matter is treated in the next problem. The crux of the argument is, that if $F(t) = 1$, we can group together several of the inter-renewal times so that the cumulative distribution function of their sum is less than 1.

Problem 4. (Optional.) If $F(t) = 1$, then since the distribution is continuous there exists some least t_0 such that $F(t_0) = 1$. Suppose r is any integer $> t/t_0$. Show that $\sum_1^r X_i > t$ provided $t/r < X_i < t_0$ $(i=1,2, \ldots, r)$ and deduce that $Pr(\sum_1^r X_i \leqslant t) < 1$. Show further that $Pr(X_1 + X_2 \ldots + X_{jr} \leqslant t) \leqslant [F_r(t)]^j$, $j = 1,2, \ldots$ and hence that $\sum_1^\infty F_n(t)$ must converge. ∎

4.5 USE OF LAPLACE TRANSFORMS

In Example 2 we were able to evaluate $\sum_1^\infty (\lambda x)^{2n-1}/(2n-1)!$ without undue difficulty. If the inter-renewal times have, say, the $\Gamma(3,\lambda)$ distribution, then we must compute $\sum_1^\infty (\lambda x)^{3n-1}/(3n-1)!$. A method for tackling the problem is outlined in the next (optional) problem.

Problem 5. Suppose $g(x) = a_0 + a_1 x + a_2 x^2 \ldots$. To evaluate $\sum_1^\infty a_{3n-1} x^{3n-1}$ we exploit the fact that if $1, \omega, \omega^2$ are the cube roots of unity, then $1 + \omega + \omega^2 = 0$. Find $g(x) + \omega g(\omega x) + \omega^2 g(\omega^2 x)$ and hence show that $\sum_1^\infty x^{3n-1}/(3n-1)! = [\exp(x) + \omega \exp(\omega x) + \omega^2 \exp(\omega^2 x)]/3$. ∎

However, the point just discussed and a great many other difficulties are most readily overcome through the use of Laplace transforms. Readers unfamiliar with this tool should study the appendix. We state two applications which are of particular use in the subject area of this chapter.

Suppose X is a non-negative continuous random variable with probability density function $f(.)$ and cumulative distribution function $F(.)$. Then denoting the Laplace transform of a function with an asterisk, by definition

$$F^*(s) = \int_0^\infty F(x)e^{-sx}dx,$$

$$= \frac{-F(x)e^{-sx}}{s}\Bigg]_0^\infty + \frac{1}{s}\int_0^\infty f(x)e^{-sx}dx$$

$$= \frac{F(0)}{s} + \frac{1}{s}f^*(s)$$

$$= \frac{1}{s}f^*(s) \quad \text{since } F(0) = 0, \text{ for a continuous distribution.}$$

Moreover, if $X_1, X_2, \ldots, X_n$ is a random sample, from the distribution of X and $T_n = \sum_1^n X_i$ has probability density function $f_n(t)$, then

$$
\begin{aligned}
f_n^*(s) &= \int_0^\infty f_n(t) e^{-st} dt \\
&= E\left(e^{-sT_n}\right) \\
&= E\left[\exp(-\sum_1^n X_i s)\right] \\
&= \prod_1^n E[\exp(-sX_i)] \\
&= \prod_1^n \int_0^\infty e^{-sx} f(x) dx \\
&= \prod_1^n f^*(s) \\
&= [f^*(s)]^n.
\end{aligned}
\tag{4.7}
$$

Hence also,

$$
\begin{aligned}
F_n^*(s) &= \frac{1}{s} f_n^*(s) \\
&= \frac{1}{s} [f_n^*(s)]^n.
\end{aligned}
\tag{4.8}
$$

We are now able to find the Laplace transform of $E(N_t) = \sum_1^\infty F_n(t)$, for

$$
\begin{aligned}
\int_0^\infty \sum_1^\infty F_n(t) e^{-st} dt &= \sum_1^\infty \int_0^\infty F_n(t) e^{-st} dt \\
&= \sum_1^\infty \frac{1}{s} [f^*(s)]^n, \text{ using equation (4.8)}, \\
&= \frac{f^*(s)}{s[1-f^*(s)]}, \text{ since } f^*(s) < 1.
\end{aligned}
\tag{4.9}
$$

Having found the Laplace transform of $E(N_t)$, using equation (4.9), we then attempt to identify (by inspection) the form of $E(N_t)$ itself. This assumes that Laplace transforms have unique inverses.

Example 3
If X_i has p.d.f. $f(x_i) = \lambda(\lambda x_i) \exp(-\lambda x_i)$, then $f^*(s) = [\lambda/(\lambda+s)]^2$, hence the Laplace transform of $E(N_t)$ is

$$
\frac{1}{s} \frac{[\lambda/(\lambda+s)]^2}{\{1-[\lambda/(\lambda+s)]^2\}}
$$

$$= \frac{1}{s^2} \cdot \frac{\lambda^2}{(s+2\lambda)}$$

$$= \frac{\lambda}{2s^2} - \frac{1}{4s} + \frac{1}{4(s+2\lambda)} \cdot$$

Now the functions which have Laplace transforms $1/(s+2\lambda)$, $1/s$, $1/s^2$ are $\exp(-2\lambda t)$, 1, t respectively. Hence,

$$E(N_t) = \frac{\lambda t}{2} - \frac{1}{4}[1-\exp(-2\lambda t)]$$

as already obtained in Example 2.

Problem 6. In an ordinary renewal process, the time to failure of a component is distributed as the sum of two independent variables having exponential distributions with parameters λ_1, λ_2 respectively. Use the method of Laplace transformation to show that the expected number of renewals in $(0,t)$ is

$$\left(\frac{\lambda_1 \lambda_2}{\lambda_1 + \lambda_2}\right) t - \frac{\lambda_1 \lambda_2}{(\lambda_1 + \lambda_2)^2} + \frac{\lambda_1 \lambda_2}{(\lambda_1 + \lambda_2)^2} \exp[-(\lambda_1 + \lambda_2)t]. \qquad \blacksquare$$

Problem 7. In an ordinary renewal process, the p.d.f. of the life of an individual component is $\frac{1}{2}\exp(-x) + \exp(-2x)$, $x > 0$. Show that the Laplace transform of the expected number of renewals in $(0,t)$ is

$$\frac{4}{3s^2} + \frac{1}{9s} - \frac{2}{9(2s+3)} \cdot$$

and hence that $E(N_t) = \frac{4t}{3} + \frac{1}{9} - \frac{1}{9}\exp(-3t/2). \qquad \blacksquare$

4.6 THE RENEWAL FUNCTION

The expected number of renewals in the interval $(0,t)$, regarded as a function of t, is called the **renewal function** and denoted by $H(t)$. The general importance of this function lies in the fact that it uniquely determines the renewal process. The proof of this result is beyond the scope of this text, but we may exploit it to verify that a claimed renewal function is indeed correct. To this end we first obtain an identity satisfied by $H(t)$ for an ordinary renewal process.

We split up the number of renewals in $(0,t)$, if any, into the first at time x and the remainder in time $t-x$. That is to say

$$N_t = 1 + N_{t-x}, \quad x \leq t,$$

$$N_t = 0 \qquad , \quad x > t.$$

Hence $H(t|x) = E(N_t|x) = 1 + H(t-x|x), \ x \leqslant t,$

$\qquad H(t|x) = 0 \ \text{if} \ x > t.$

But the time to the first renewal has p.d.f. $f(x)$, hence

$$H(t) = E[H(t|x)] = \int_0^\infty H(t|x)f(x)dx$$

$$= \int_0^t [1 + H(t-x)] \ f(x)dx,$$

i.e.

$$H(t) = \int_0^t H(t-x)f(x)dx + F(t). \qquad (4.10)$$

The result in (4.10) is known as the renewal equation.

Example 4

In Example 3, we have $f(x) = \lambda(\lambda x) \exp(-\lambda x)$ and $H(t) = \dfrac{\lambda t}{2} - \dfrac{1}{4} [1-\exp(-2\lambda t)]$.

Thus $F(t) = \displaystyle\int_0^t f(x)dx$ and $H(t-x) = \dfrac{\lambda(t-x)}{2} - \dfrac{1}{4} \{1-\exp[-2\lambda(t-x)]\}$. These

expressions maybe used to verify (4.10). The computations are routine and need not detain us.

Problem 8. Show that the $H(t)$ in Problem 7, satisfies equation (4.10). ■

Problem 9. Take the Laplace transform of both sides of equation (4.10) and hence find $H^*(s)$. (Hint: change the order of integration in the double integral.) ■

* Problem 10. (More difficult.) Equation (4.10) is a particular case of the type of equation

$$K(t) = k(t) + \int_0^t K(t-x)f(x)dx$$

where $f(x) = 0$ for $x < 0$, $k(t) = 0$ for $t < 0$, and $k(.)$ is bounded. Verify by substitution that one solution is

$$K(t) = \int_0^t k(t-x) \sum_1^\infty f_n(x)dx + k(t),$$

where $f_n(.)$ is the n-fold convolution of $f(.)$ with itself.

(Note: $f_{n+1}(x) = \displaystyle\int_0^x f_n(x-y)f(y)dy$.) ■

Convolution Formulae

It is frequently convenient to find the distribution function of T_n from the fact that $T_n = T_{n-1} + X_n$. Thus, for an ordinary renewal process,

$$Pr(T_n \leqslant t) = \int_0^t Pr(T_n \leqslant t | X_1 = x_1) f(x_1) dx_1 .$$

But

$$Pr(T_n \leqslant t | X_1 = x_1) = Pr \left[\sum_1^n X_i \leqslant t | X_1 = x_1 \right]$$

$$= Pr \left[\sum_2^n X_i \leqslant t - x_1 | X_1 = x_1 \right], x_1 \leqslant t$$

$$= Pr \left[\sum_2^n X_i \leqslant t - x_1 \right], \text{ from the mutual independence.}$$

Hence, in terms of the relevant distribution functions,

$$F_n(t) = \int_0^t F_{n-1}(t - x_1) f(x_1) dx_1, \quad n \geqslant 2. \tag{4.11}$$

The integral in equation (4.11) is the convolution between $F_{n-1}(.)$ and $f(.)$. The result in equation (4.11) may be used to obtain equation (4.10). The reader should verify this by using $H(t) = \sum_1^\infty F_n(t)$ and using (4.11) in the right-hand side of (4.10).

Problem 11. Confirm equation (4.11) by taking the Laplace transform of both sides. ■

Problem 12. Let $X_2, X_3, \ldots X_n$ be independent, non-negative and continuous random variables, and let $\sum_2^n X_i$ have cumulative distribution function $F_{n-1}(.)$. Let X_1, though independent of $X_i (i \geqslant 2)$, have probability density function $g(.)$. Use a direct argument to show that

$$Pr \left[\sum_1^n X_i \leqslant t \right] = \int_0^t F_{n-1}(t - x_1) g(x_1) dx_1 \tag{4.12}$$

$$= \int_0^t F_{n-1}(x_1) g(t - x_1) dx_1 .$$

Hence, or otherwise find the probability density function of $\sum_1^n X_i$. ■

Problem 13. For an ordinary renewal process, show that

(i) $\displaystyle\int_0^t H(t-x) \sum_1^n f_n(x)\mathrm{d}x = \sum_2^\infty (n-1)F_n(t)$

(ii)   variance $(N_t) = 2 \displaystyle\int_0^t H(t-x) \sum_1^\infty f_n(x)\mathrm{d}x + H(t) - [H(t)]^2$.   ∎

* 4.7 APPROXIMATING THE RENEWAL FUNCTION

We have already shown that for an ordinary renewal process for which $F(t) < 1$, the renewal function, $H(t)$, satisfies

$$F(t) \leqslant H(t) \leqslant F(t) / [1 - F(t)]. \tag{4.13}$$

We have also remarked that $H(t)$ is not, in general, proportional to t. The next problem shows that if $H(t)$ is proportional to t, then the distribution of the inter-renewal times must be exponential.

Problem 14. Starting with the renewal equation (4.10),

$$H(t) = F(t) + \int_0^t H(t-x)f(x)\mathrm{d}x,$$

show that if $H(t) = t/\mu$ then, after substituting and then differentiating with respect to t.

$$\frac{1}{\mu} = f(t) + \frac{1}{\mu}F(t).$$

Deduce that $F(t) = 1 - \exp(-t/\mu)$.   ∎

Nevertheless, the reader may well be thinking that 'for large t, the expected number of renewals must be nearly proportional to t', and in fact, $\lim_{t \to \infty} [H(t)/t]$ is indeed $1/\mu$ where μ is the mean life of each component.

A proof of this result may be found in Ross [1]. We discuss the weaker, and rather easier result, that

$$H(t) \geqslant \frac{t}{\mu} - 1. \tag{4.14}$$

Let T_N be the time to the first renewal after the fixed time t, then, by definition,

$$T_N \geqslant t,$$

then certainly

$$E(T_N) \geqslant t.$$

But N is one more than the number of renewals in $(0,t)$, hence

$$H(t) = E(N) - 1.$$

From a result due to Wald,

$$E(T_N) = E(N)E(X), \qquad (4.15)$$

hence

$$H(t) = \frac{E(T_N)}{\mu} - 1 \geqslant \frac{t}{\mu} - 1.$$

The result in (4.15) is more subtle than it looks, since N is not independent of T_N. For a proof see Barlow and Proschan [2]. We state an even better result. If $\{T_n\}$ is an ordinary renewal process such that $T_n = \sum_{1}^{n} X_i$, $E(X_i) = \mu$, $V(X_i) = \sigma^2$, then

$$H(t) = \frac{t}{\mu} + \frac{\sigma^2 - \mu^2}{2\mu^2} + o(1), \qquad (4.16)$$

where $o(1)$ signifies a term tending to zero as t tends to infinity. For instance in Example 2, we have $E(X_i) = 2/\lambda$, $V(X_i) = 2/\lambda^2$ and it is readily checked that $H(t)$ is of the form (4.16).

Problem 15. Verify formula (4.16) for Problems 6 and 7. ∎

A rigorous proof of (4.16) is rather difficult, but the next problem will go some way towards making it seem plausible. We consider the Laplace transform of $H(t)$. Now

$$H^*(s) = \int_0^\infty e^{-st} H(t) dt,$$

and $\exp(-st)$ is negligible for large values of t, except when s is small. This suggests that the behaviour of $H(t)$ as t tends to infinity is related to the behaviour of $H^*(s)$ as s tends to zero. The Laplace transform of (4.16) is

$$H^*(s) = \frac{1}{s^2 \mu} + \frac{1}{s} \left(\frac{\sigma^2 - \mu^2}{2\mu^2} \right) + o\left(\frac{1}{s} \right). \qquad (4.17)$$

But from (4.9) we have that, for the ordinary renewal process,

$$H^*(s) = \frac{1}{s} \frac{f^*(s)}{[1 - f^*(s)]}, \qquad (4.18)$$

where $f^*(s)$ is the Laplace transform of the p.d.f. of the inter-renewal times. In particular cases it can be shown directly that for the $f^*(s)$ in question, (4.18), is of the form (4.17), and hence that (4.16) holds. Cox [3] has given a fuller discussion and several illustrative examples.

Problem 16. If $G^*(s)$ is the Laplace transform of $G(t)$, then from Appendix 2, Example 11

$$\lim_{s \to 0} [s\{G^*(s)\}] = \lim_{t \to \infty} [G(t)].$$

This result provides one way of exploring the behaviour of $G(t)$ for large values of t.

Let $H(t)$ be the renewal function of an ordinary renewal process for which the inter-renewal times have mean μ and variance σ^2. If $G(t) = H(t) - t/\mu$, show that

$$\lim_{t \to \infty} [G(t)] = (\sigma^2 - \mu^2)/2\mu^2. \quad \blacksquare$$

4.8 MODIFIED RENEWAL PROCESS

In a **modified renewal process**, the first component in use may have a different distribution of time to failure from that of subsequent components. This will be the case, if the first component is not new. The corresponding measures will be given an additional m, while for an ordinary process, we use o. Many of the formulae obtained from an ordinary process require little change. Thus, if $N_t^{(m)}$ is the number of renewals in $(0,t)$ and $T_n^{(m)}$ the time to the nth renewal in a modified process, then it remains true that

$$Pr(N_t^{(m)} \geqslant n) = Pr(T_n^{(m)} \leqslant t). \tag{4.19}$$

Hence,

$$H_m(t) = E(N_t^{(m)}) = \sum_1^\infty F_n^{(m)}(t), \tag{4.20}$$

where $F_n^{(m)}(t)$ is the cumulative distribution function of $T_n^{(m)}$. If $T_n^{(m)}$ consists of a first component with p.d.f. $g(.)$ and $(n-1)$ components each with p.d.f. $f(.)$, then

$$F_n^{(m)}(t) = \int_0^t F_{n-1}(t-x)g(x)dx. \tag{4.21}$$

Problem 17. Show that the Laplace transform of $f_n^{(m)}(t)$ is

$$g^*(s) \, [f^*(s)]^{n-1} \tag{4.22}$$

and that

$$H_m^*(s) = \frac{g^*(s)}{s[1-f^*(s)]} \tag{4.23} \quad \blacksquare$$

Problem 18. Use equation (4.20), to show that

$$H_m(t) = G(t) + \int_0^t H_m(t-x)f(x)dx, \qquad (4.24)$$

where $G(.)$ is the c.d.f. of the first component. The life of a spark plug has an exponential distribution with parameter λ_1. When a plug fails, the time it takes to replace it with a new one has an exponential distribution with parameter $\lambda_2 (\neq\lambda_1)$. All lifetimes and replacement times are to be considered as independent random variables. If the initial plug is new, show that the renewal function for the number of failures in $(0,t)$ is

$$H_m(t) = \frac{\lambda_1\lambda_2 t}{\lambda_1+\lambda_2} + \frac{\lambda_1^2}{(\lambda_1+\lambda_2)^2}[1-\exp\{-(\lambda_1+\lambda_2)t\}],$$

(i)   by verifying that $H_m(t)$ satisfies equation (4.24) and
(ii) by computing $H_m^*(s)$.
(Note that the process is 'modified' since the time to the first failure has a distribution differing from that between subsequent failures.)   ∎

We have seen that the only ordinary renewal process for which $H(t) = t/\mu$ for all t arises when the individual components have exponentially distributed lives. Is there perhaps a modified renewal process such that $H_m(t) = t/\mu$? If there is, then its Laplace transform must be $1/(\mu s^2)$. But then, from the result obtained in Problem 17, equation (4.23),

$$\frac{g^*(s)}{s[1-f^*(s)]} = \frac{1}{\mu s^2};$$

that is

$$g^*(s) = \frac{1}{\mu s} - \frac{f^*(s)}{\mu s}$$

or

$$g(t) = \frac{1}{\mu}[1-F(t)]. \qquad 4.25)$$

The particular modified process for which the density of the initial component satisfies equation (4.25), is further qualified as an **equilibrium renewal process**. Partial support for using this description comes from the limiting behaviour of the renewal function for *any* ordinary renewal process.

Problem 19. From the identity (4.24) in Problem 18, assume that $H_m(t) = t/\mu$ and hence show that

$$G(t) = \frac{1}{\mu}\int_0^t [1-F(x)]\ dx$$

and recover equation (4.25).   ∎

Renewal Density

In an ordinary renewal process, the probability that the first failure take place in the small interval $(t, t+\delta t)$ is

$$\int_t^{t+\delta t} f(x)dx.$$

Since the probability of more than one failure in $(t, t+\delta t)$ is $o(\delta t)$, the probability that *some* failure takes place in $(t, t+\delta t)$ is

$$\sum_1^\infty \int_t^{t+\delta t} f_n(x)dx + o(\delta t),$$

$$= \int_t^{t+\delta t} \sum_1^\infty f_n(x)dx + o(\delta t),$$

$$= \sum_1^\infty f_n(t)\,\delta t + o(\delta t). \tag{4.26}$$

The limit of the ratio of the quantity in (4.26) to δt as $\delta t \to 0$, is $\sum_1^\infty f_n(t)$ and is appropriately called the **renewal density**, $h_o(t)$. Since $H_o(t) = \sum_1^\infty F_n(t)$, it is apparent that $h_o(t) = H_o'(t)$. It is usually more convenient to compute $h_o(t)$ and then find $H_o(t)$ by integration.

Problem 20. For an ordinary renewal process use $h_o(t) = \sum_1^\infty f_n(t)$ to show that

$$h_o(t) = f(t) + \int_0^t h_o(t-x)f(x)dx. \tag{4.27}$$

obtain the same result by differentiating both sides of equation (4.10). Show that

$$h_o^*(s) = f^*(s)/[1-f^*(s)]. \tag{4.28} \quad \blacksquare$$

Problem 21. In an ordinary renewal process, show that if the inter-renewal times have independent $\Gamma(4,\lambda)$ distributions, then the renewal density is

$$h_o(t) = \frac{\lambda}{4}[1-\exp(-2\lambda t)-2\sin(\lambda t)\exp(-\lambda t)]. \quad \blacksquare$$

Problem 22. In a modified renewal process, show that

$$h_m(t) = g(t) + \int_0^t h_m(t-x)f(x)dx. \tag{4.29} \quad \blacksquare$$

4.9 BACKWARD RECURRENCE TIME

Suppose we examine a modified renewal process at time t and enquire as to the distribution of the age, U, of the part functioning at time t. We shall find that the renewal density, $h_m(.)$ plays here a useful role. Now it may be that no renewal has taken place. This will be the case if the very first unit survives for longer than t – which it does with probability $1-G(t)$. We then take $u = t$. On the other hand, a last renewal can have taken place in the interval $(t-u-\delta u, t-u)$, with probability $h_m(t-u-\delta u)\delta u + o(\delta u)$ and the new unit there inserted lasted at least a further $u+\delta u$, which it does with probability $1-F(u+\delta u)$. So, if $\phi_m(u)$ is the p.d.f. of U, then

$$\phi_m(u)\delta u + o(\delta u) = h_m(t-u-\delta u) \left[1-F(u+\delta u)\right] \delta u + o(\delta u).$$

Hence, in the limit as $\delta u \to 0$,

$$\phi_m(u) = h_m(t-u) \left[1-F(u)\right], 0 \leqslant u < t. \tag{4.30}$$

The distribution of U is mixed. The continuous part, is as in equation (4.30), the discrete part has probability $1-G(t) = \Phi_m(t)$.

Problem 23. Verify that

$$\Phi(t) + \int_0^t h_m(t-u) \left[1-F(u)\right] \, du = 1. \quad \blacksquare$$

Problem 24. Show that, since its distribution is mixed, the mean of the backward recurrence time is

$$E(U) = t\Phi_m(t) + \int_0^t u h_m(t-u) \left[1-F(u)\right] \, du.$$

In an ordinary renewal process, the inter-renewal times are exponentially distributed with parameter λ. Show that $E(U) = \left[1-\exp(-\lambda t)\right]/\lambda$. $\quad \blacksquare$

As t increases, $G(t) \to 1$, hence $\Phi(t) \to 0$ and the 'influence of the origin' wears off. The behaviour of $\phi_m(u)$ as t increases, where $u < t$, is not so unambiguous. For fixed u, $h_m(t-u) \to 1/\mu$. This suggests the limiting p.d.f. of U is $\phi_m(u) = \left[1-F(u)\right]/\mu$.

Problem 25. In the equilibrium renewal process, show that

$$\Phi_e(t) = 1 - \frac{1}{\mu} \int_0^t \left[1-F(x)\right] dx,$$

$$\phi_e(u) = \frac{1}{\mu} \left[1-F(u)\right], 0 \leqslant u < t.$$

4.10 FORWARD RECURRENCE TIME

In the same way, at a fixed time t, the *forward* recurrence time, V, is defined as the time from t to the next renewal. That is to say, V is the remaining lifetime of the component in use at time t. Suppose the next renewal takes place in $(v, v+\delta v)$ *after* time t. There are two possibilities. The first component (possibly not new) may fail in $(t+v, t+v+\delta v)$ and this happens with probability $g(v+t)\,\delta v + o(\delta v)$. Alternatively, there was some *last* renewal at time $y (< t)$ and the (new) part then inserted fails in $(t-y+v, t-y+v+\delta v)$ later. Hence the p.d.f. $\psi_m(v)$, of V after integrating out y is

$$\psi_m(v) = g(v+t) + \int_0^t h_m(y)f(t+v-y)\mathrm{d}y. \qquad (4.31)$$

Problem 26. Verify that in the ordinary renewal process, for which $f(x) = \lambda\exp(-\lambda x)$, $\psi_0(v) = \lambda\exp(-\lambda v)$ (which should come as no surprise). ∎

Problem 27. In an ordinary renewal process, the inter-renewal times have density $f(x) = \lambda^2 x \exp(-\lambda x)$. Show that

$$h_0(y) = \lambda[1-\exp(-2\lambda y)]/2.$$

Hence, show that the density of the forward renewal time is

$$\psi_0(v) = \frac{\lambda^2}{2}\exp(-\lambda v)\,[v\{1+\exp(-2\lambda t)\}] + \frac{1}{\lambda} - \frac{1}{\lambda}\exp(-2\lambda t).$$

(Note: it is always worth checking that, when $t = 0$, $\psi_0(v)$ collapses to the p.d.f. for the life of the first part, viz. $g(v)$.) ∎

Problem 28. In an ordinary renewal process, show that the probability that the forward recurrence time exceeds the backward recurrence time, at time t, is

$$1-F(2t) + \int_0^t [1-F(2t-2x)]h_0(x)\mathrm{d}x. \quad ∎$$

Problem 29. For the equilibrium renewal process show that

$$\psi_e(v) = [1-F(v)]/\mu. \quad ∎$$

4.11 THE RELIABILITY OF AN INDIVIDUAL COMPONENT

That new components should be 'reliable' in some sense is clear. While there is no single measure of performance, it does seem reasonable that a component should have a high probability of lasting for at least some minimum acceptable time. Otherwise installation and purchasing costs may rise to an unacceptable level.

In the present context, we shall call the probability that a new component lasts at least time t, the **reliability** of the component. (The reliability of a *system* is discussed in Chapter 5.) As a function of t, we denote this probability by $\bar{F}(t)$. (In the literature, $\bar{F}(.)$ is also known as the **survivor function**, or the **hazard function**.) Evidently $\bar{F}(t) = 1 - F(t)$ where $F(.)$ is the c.d.f. of the time to failure of the component. Hence $\lim_{t \to \infty} [\bar{F}(t)] = 0$ and, for a positive random variable, $\bar{F}(0) = 1$. For a continuous random variable, $\bar{F}(t)$ is strictly decreasing.

Problem 30. Calculate $\bar{F}(t)$, when T has the $\Gamma(2,\lambda)$ distribution.   ■

The least sophisticated approach to breakdowns involves changing a component only when it fails. The cost of such emergency replacements may make some additional form of preventive maintenance desirable; that is to say, a component may also be changed before it has failed. If this is done whenever a component has functioned continuously for a fixed time then we have adopted an **age-replacement policy**. Failed components are also changed. This may be contrasted with a **block policy** under which components are changed both on failing and at fixed times, regardless of their ages. Such policies lead us to consider a generalisation of the reliability of a new component, namely, the conditional probability that a component which has functioned for a time t_0 will function for at least a further time $t_1 - t_0$ (> 0). Now,

$$Pr[T > t_1 | T > t_0] = Pr[T > t_1]/Pr[T > t_0]$$

$$= \bar{F}(t_1)/\bar{F}(t_0).$$

If at t_0, this old component is exchanged, then the new component has probability $\bar{F}(t_1 - t_0)$ of lasting for at least a time $t_1 - t_0$. We say that **new parts are better than used** (or that the distribution is N.B.U.) if and only if

$$\bar{F}(t_1 - t_0) > \bar{F}(t_1)/\bar{F}(t_0) \tag{4.32}$$

for all t_1 $(> t_0)$ and every $t_0 > 0$.

Example 5

If the time to failure of a new component has a $\Gamma(2,\lambda)$ distribution, then from Problem 30, $\bar{F}(t) = (1 + \lambda t)\exp(-\lambda t)$. Hence $\bar{F}(t_1 - t_0)\bar{F}(t_0) - \bar{F}(t_1)$ is

$$\{[1 + \lambda(t_1 - t_0)] \exp[-\lambda(t_1 - t_0)]\} [(1 + \lambda t_0) \exp(-\lambda t_0)] - (1 + \lambda t_1)\exp(-\lambda t_1)$$

$$= \exp(-\lambda t_1) [(1 + \lambda t_1 - \lambda t_0)(1 + \lambda t_0) - (1 + \lambda t_1)]$$

$$= \lambda^2 t_0(t_1 - t_0) \exp(-\lambda t_1) > 0 \text{ for } t_1 > t_0 \text{ for all } t_0 > 0.$$

Hence for this distribution, new parts are better than used.

In contrast, we say a distribution is N.W.U., **new parts are worse than used**, if and only if

$$\bar{F}(t_1 - t_0) < \bar{F}(t_1)/\bar{F}(t_0)$$

for all t_1 ($> t_0$) and every $t_0 > 0$.

Problem 31. Show that the distribution for which

$$\bar{F}(t) = (1-p)\exp(-\lambda_1 t) + p \exp(-\lambda_2 t), 0 < p < 1, \lambda_1 \neq \lambda_2,$$

is N.W.U. ∎

Problem 32. The continuous and positive random variable X is said to have the *Weibull distribution* with parameters α, λ if it has p.d.f.

$$f(x) = \alpha\lambda(\lambda x)^{\alpha-1} \exp[-(\lambda x)^\alpha], \alpha > 0, \lambda > 0.$$

Show that X is N.B.U. or N.W.U. according as $\alpha > 1$ or $\alpha < 1$. ∎

Under either an age replacement or a block policy, components may be changed before they have failed. If we match the age at which components must be changed with the duration of the intervals at which regular changes are made, then it is intuitively clear that the number of replacements under a block policy will not be less than under an age policy. For under a block policy, a component must be changed at the next regular time, however short a service it has provided. This unfavourable aspect is not conclusive, for block policies are obviously easier to administer, since we need not keep track of the ages of components.

For the age replacement policy, the failure times constitute a renewal process. Of course each failure may be preceded by a number of changes of component due to age. We examine the probability that the first failure has not occurred by time t when a component of age t_a must be changed. This probability clearly depends on the number of components changed before t. Suppose then that $kt_a < t < (k+1)t_a$. No failure has been observed by t if k components independently survive for at least time t_a and the $(k+1)$th component lasts for at least $t - kt_a$. This event has probability

$$[\bar{F}(t_a)]^k \bar{F}(t - kt_a). \tag{4.33}$$

Example 6
If the life of each component is exponentially distributed with parameter λ, then (4.33) becomes $[\exp(-\lambda t_a)]^k \exp[-\lambda(t - kt_a)] = \exp(-\lambda t)$, which should cause the reader no surprise.

We can use (4.33) to obtain the expected number of age replacements before the first replacement due to failure. The probability that there are at *least* k age replacements is $[\bar{F}(t_a)]^k$. Hence the expected number of age replacements is

$$\sum_{k=1}^{\infty} [\bar{F}(t_a)]^k = \bar{F}(t_a)/F(t_a).$$

(Note the calculation of $E(K)$ as $\Sigma \, Pr(K \geqslant k)$. See Appendix 3.)

Problem 33. Show that the expected *time* to the first failure in an age replacement policy is

$$\frac{1}{F(t_a)} \int_0^{t_a} \bar{F}(y)\mathrm{d}y. \quad \blacksquare$$

Additional properties of age-replacement policies can be deduced for particular types of component. Thus for N.B.U. components consider again the probability of no failure by time t. This is $[\bar{F}(t_a)]^k \, \bar{F}(t-kt_a)$ for $kt_a < t < (k+1)t_a$. We may write this probability in the form $[\bar{F}(t_a)]^{k-1} \, \bar{F}(t_a)\bar{F}(t-kt_a) > [\bar{F}(t_a)]^{k-1} \, \bar{F}[t-(k-1)t_a]$, since the distribution is N.B.U. This confirms that the kth age replacement is worthwhile, in terms of probability of survival. Indeed by continued application,

$$[\bar{F}(t_a)]^k \, \bar{F}(t-kt_a) > \bar{F}(t) = Pr[\text{first component lasts at least time } t].$$

We are now moved to consider whether a reduction in the replacement age inevitably increases the probability of survival of the system. The N.B.U. property is not quite strong enough to guarantee this. However, any t_a is better than mt_a where m is an integer. This is the subject of Problem 34 (optional).

Problem 34. In a replacement policy at age t_a, suppose $kt_a < t < (k+1)t_a$. Let m be an integer and ℓ the greatest integer such that $m\ell \leqslant k$. Show that for N.B.U. components

$$\bar{F}(mt_a) < [\bar{F}(t_a)]^m.$$

Hence, or otherwise, deduce that

$$[\bar{F}(t_a)]^k \, \bar{F}(t-kt_a) > [\bar{F}(mt_a)]^\ell \, \bar{F}(t-m\ell t_a), \quad kt_a < t < (k+1)t_a. \quad \blacksquare$$

4.12 INCREASING FAILURE RATE

For the stronger property needed to compare any two replacement ages, we reconsider the reliability of a component having age t_0. The conditional probability that this component lasts until at least t_1, given that it has lasted for at least t_0 ($< t_1$), is $\bar{F}(t_1)/\bar{F}(t_0)$. If for each fixed $t_1-t_0 = \tau$, this is a decreasing function of t_0, then we say that the distribution of the life of the component has an **increasing failure rate** (I.F.R.). More bluntly, the component is ageing. This property implies that the component is N.B.U., for

$$\frac{\bar{F}(t_1 - t_0)}{\bar{F}(0)} > \frac{\bar{F}(t_1)}{\bar{F}(t_0)} \ ,$$

and is equivalent to (4.32) since $\bar{F}(0) = 1$.

Since $\bar{F}(t_0 + \tau)/\bar{F}(t_0)$ is decreasing in t_0, for a continuous distribution its derivative with respect to t_0 is negative. That is,

$$[-\bar{F}(t_0)f(t_0 + \tau) + \bar{F}(t_0 + \tau)f(t_0)] / [\bar{F}(t_0)]^2 < 0,$$

which implies

$$\frac{f(t_0)}{\bar{F}(t_0)} < \frac{f(t_0 + \tau)}{\bar{F}(t_0 + \tau)} \ .$$

Indeed it is not unusual for the concept of an ageing component to be expressed directly in terms of $r(t) = f(t)/\bar{F}(t)$, where $r(t)$ is called the **failure rate**. Thus a component is said to have an **increasing failure rate** (I.F.R.) if and only if $r(t)$ is increasing in t. If $r(t)$ is decreasing in t, then the component is said to have a **decreasing failure rate** (D.F.R.). $r(t)$ attracts the name 'failure rate' in virtue of the property that the probability that a component, known to be functioning at time t, fails in the interval $(t, t+\delta t)$ is $r(t)\,\delta t + o(\delta t)$.

Example 7
Consider again the distribution for which

$$f(t) = (1-p)\lambda_1 \exp(-\lambda_1 t) + p\lambda_2 \exp(-\lambda_2 t), t > 0, 0 < p < 1.$$

Then

$$r(t) = \lambda_2 + \{(1-p)(\lambda_1 - \lambda_2)/[1 - p + p\exp(\lambda_1 - \lambda_2)t]\}$$

and is *decreasing* in t unless $\lambda_1 = \lambda_2$.

Problem 35. Show that for the $\Gamma(\alpha, \lambda)$ distribution, $r(t)$ increases with t if $\alpha > 1$ and decreases with t if $\alpha < 1$. Note that if $\alpha = 1$, $r(t) = \lambda$, a constant. The exponential distribution is neither I.F.R. nor D.F.R., though a limiting case for both. ∎

Problem 36. Starting with $r(t) = f(t)/\bar{F}(t)$, show that $\bar{F}(t) = \exp[-\int_0^t r(y)\mathrm{d}y]$.

Hence show that if $r(t)$ is increasing in t, then $\bar{F}(t+x)/\bar{F}(t)$ is decreasing in t. ∎

We now return briefly to the question of comparing replacement policies for two different replacement ages t_1, t_2. In particular, we consider the difference in the probabilities of no failure by time t. Suppose that $n_1 t_1 < t < (n_1 + 1)t_1$ and that $n_2 t_2 < t < (n_2 + 1)t_2$. If $t_1 < t_2$, then $n_1 \geqslant n_2$, n_1 and n_2 being integers. Then the difference in the probabilities is

$$[\bar{F}(t_1)]^{n_1}\,\bar{F}(t - n_1 t_1) - [\bar{F}(t_2)]^{n_2}\,\bar{F}(t - n_2 t_2). \tag{4.34}$$

Since

$$[\bar{F}(t_1)]^{n_1}\, \bar{F}(t-n_1 t_1) > [\bar{F}(t_1)]^{n_1-1}\, \bar{F}[t-(n_1-1)t_1] > \ldots$$
$$> [\bar{F}(t_1)]^{n_1-(n_1-n_2)}\, \bar{F}[t-(n_1-n_1+n_2)t_1]$$
$$= [\bar{F}(t_1)]^{n_2}\, \bar{F}[t-n_2 t_1],$$

the difference in (4.34) is greater than

$$[\bar{F}(t_1)]^{n_2}\, \bar{F}(t-n_2 t_1) - [\bar{F}(t_2)]^{n_2}\, \bar{F}(t-n_2 t_2). \qquad (4.35)$$

Now suppose the distribution is I.F.R. The reader should check that $[\bar{F}(x)]^k \bar{F}(t-kx)$ is decreasing in x, $kx < t < (k+1)x$. (Hint: verify that its derivative with respect to x is negative.) Hence (4.35) is positive since $t_1 < t_2$. The fact that changing components at earlier ages probably defers the evil of an actual failure is not however decisive. For such a policy may be inhibited by the increased costs involved.

A weaker comparison of new and used parts can be made in terms of *expected* remaining lives. We say that a new part is **better than used in expectation** (N.B.U.E.) if the distribution of its life, T, satisfies

$$E(T) > E(T-t\,|\,T \geqslant t), \text{ for all } t > 0. \qquad (4.36)$$

Equivalently, since $E(T) = \displaystyle\int_0^\infty \bar{F}(x)dx$, and

$$E(T-t\,|\,T \geqslant t) = \int_t^\infty \frac{\bar{F}(x)}{\bar{F}(t)}\, dx,$$

if

$$\int_0^\infty \bar{F}(x)dx > \int_t^\infty \frac{\bar{F}(x)}{\bar{F}(t)}\, dx$$
$$= \int_0^\infty \frac{\bar{F}(t+x)}{\bar{F}(t)}\, dx, \text{ for all } t. \qquad (4.37)$$

Now if a part is N.B.U. then $\bar{F}(t+x)/\bar{F}(t) < \bar{F}(x)$ for all t. Hence, by comparing the integrands, if a part is N.B.U., then it is also N.B.U.E.

Problem 37. Show that if a distribution is N.B.U.E. and has mean μ then

$$\frac{1}{\mu} \int_0^t \bar{F}(x)dx \geqslant F(t).$$

Show further, that for an ordinary renewal process with N.B.U.E. components, that the renewal function, $H_0(t)$, satisfies $H_0(t) \leqslant t/\mu$. (Hint: for an equilibrium renewal process, $H_e(t) = t/\mu$, consider Problem 19.) ∎

4.13 REPLACEMENT SCHEMES

In our discussion of the relative merits of used and new parts, we have rather lost sight of the costs which may be involved in a long run of a process which may involve many replacements. The simplest scheme is that in which a part is only exchanged when it fails. This is called a service change and, if it costs c_s per change, then, for an ordinary renewal process, the expected cost up to time t is $c_s H_o(t)$. The cost per unit time is $c_s H_o(t)/t$ and this tends to c_s/μ, as t increases, where μ is the mean life of an individual component. We call this scheme I. Scheme II is based on a block policy so that in addition to service replacements, a planned change is made at fixed times $t_0, 2t_0, \ldots$, irrespective of the age of the component then functioning and these cost c_p each. The cost of the planned replacements over $(0, kt_0]$ is kc_p. As for the service replacements, since a new part is inserted at $it_0 (i = 1, 2, \ldots, k)$ we have effectively an ordinary renewal process for each of k intervals of duration t_0. The expected cost for service replacements for each of these is $H_o(t_0)$. Hence the expected cost per unit time for scheme II is

$$[kc_p + c_s k H_o(t_0)]/kt_0 = [c_p + c_s H_o(t_0)]/t_0.$$

Problem 38. Compute the expected costs per unit time over the interval $(0, kt_0)$ for schemes I and II when the life of a part has a $\Gamma(2, \lambda)$ distribution. ∎

Problem 39. Simplified comparisons for schemes I and II can be made via the approximation in equation (4.16),

$$H_o(t) = \frac{t}{\mu} + \frac{\sigma^2 - \mu^2}{2\mu^2} + o(1).$$

By considering the limit as k increases, show that the expected cost per unit time over $(0, kt_0)$ is less for scheme II if

$$\frac{c_p}{c_s} < \left(\frac{\mu^2 - \sigma^2}{2\mu^2}\right), \text{ when } \sigma < \mu. ∎$$

An objection to scheme II is that block replacement may throw away components which are too 'young'. In scheme III, a failed component is replaced at cost c_s per unit and a unit is also replaced at critical age t_a at c_a per unit. We can find the approximate cost of this scheme over a long interval $(0, t)$. Since $\bar{F}(t_a)$ is the probability that a component endures for at least time t_a, it is also the proportion of changes costing c_a. Similarly $F(t_a)$ is the proportion costing c_s. But, for large t, the expected number of changed components (of either kind) is approximately

$$t/ \text{[mean time between changes]}.$$

The mean time between changes is

$$\int_0^{t_a} xf(x)dx + t_a \bar{F}(t_a) = \int_0^{t_a} \bar{F}(x)dx.$$

Hence, the expected cost per unit time for scheme III is approximately

$$\frac{c_s F(t_a) + c_a \bar{F}(t_a)}{\int_0^{t_a} \bar{F}(x)dx}.$$

Problem 40. Recover the expected cost per unit time for scheme III by considering the expected time between service failures and the expected number of age-replacements between service failures. ∎

4.14 OPTIMAL CHOICE OF INTERVAL

We have already remarked that for I.F.R. components, a shorter replacement age decreases the chance of a failure before a fixed time, but drew attention to the fact that we must take account of the costs involved. For scheme III, we may search for the choice of t_a which minimises the expected cost per unit time. The derivative of $[c_s F(t_a) + c_a \bar{F}(t_a)] / \int_0^{t_a} \bar{F}(x)dx$ with respect to t_a is

$$\frac{(c_s - c_a)\bar{F}(t_a)\left[r(t_a) \int_0^{t_a} \bar{F}(x)dx - F(t_a) - c_a/(c_s - c_a)\right]}{\left[\int_0^{t_a} \bar{F}(x)dx\right]^2}.$$

To find a minimum, we need to consider the solution of

$$r(t_a) \int_0^{t_a} \bar{F}(x)dx - F(t_a) = c_a/(c_s - c_a),$$

where of course $c_a < c_s$ if age-replacements are to be worthwhile at all! For I.F.R. components, the left-hand side is increasing. Hence there is at most one solution. If there is no solution, age replacements should not be made.

Problem 41. Show for scheme III, that as $t_a \to \infty$, the expected cost per unit time tends to c_s/μ, where μ is the mean life of a component. Show further that for I.F.R. components, age-replacements should never be made before age $c_a\mu/c_s$. ∎

Problem 42. The expected cost per unit time for scheme II is $[c_p + c_s H(t_0)]/t_0$, where t_0 is the (fixed) interval between block replacements. Find the optimal choice of t_0 when components have $\Gamma(2,\lambda)$ distributions. ∎

*4.15 BOUNDS FOR RELIABILITY FUNCTIONS

We shall shortly show that $[\bar{F}(t)]^{1/t}$ is decreasing or increasing in t according as the distribution is I.F.R. or D.F.R. Although $\bar{F}(t)$ is decreasing in t, the behaviour of $[\bar{F}(t)]^{1/t}$ or equivalently $[\log \bar{F}(t)]/t$, is not predetermined. For example, if $\bar{F}(t) = \exp(-t^\alpha)$, then $[\log \bar{F}(t)]/t = -t^{\alpha-1}$ which is decreasing if $\alpha > 1$ but increasing if $\alpha < 1$.

Now $r(t) = f(t)/[1-F(t)] = -\dfrac{\mathrm{d}}{\mathrm{d}t}[\log \bar{F}(t)]$. But if the distribution is I.F.R., then $r(t)$ is increasing and hence the derivative of $\log \bar{F}(t)$ is decreasing. A glance at Fig. 4.1. will show why, in that case, $[\log \bar{F}(t)]/t = \log [\bar{F}(t)]^{1/t}$ is also decreasing. Notice that $[\log \bar{F}(t)]/t$ is the slope of OP.

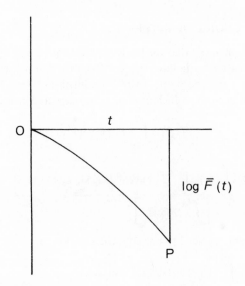

Figure 4.1

In any case,

$$
\begin{aligned}
\frac{\mathrm{d}}{\mathrm{d}t}\left[\frac{1}{t}\log \bar{F}(t)\right] &= -\frac{1}{t^2}\left[tr(t) + \log \bar{F}(t)\right] \\
&= -\frac{1}{t^2}\left[tr(t) - \int_0^t r(y)\mathrm{d}y\right] \\
&= -\frac{1}{t^2}\left[\int_0^t \{r(t) - r(y)\}\,\mathrm{d}y\right] \\
&< 0,
\end{aligned}
$$

since $r(.)$ is increasing. Similarly if the distribution in D.F.R., then $[\bar{F}(t)]^{1/t}$ is increasing.

These results are the basis of various interesting inequalities. Thus suppose t_p is the lower $100p\%$ point of the distribution. That is, $F(t_p) = p$, or $\bar{F}(t_p) = 1-p$. Then for *any* I.F.R. distribution,

$$[\bar{F}(t)]^{1/t} \geqslant [\bar{F}(t_p)]^{1/t_p} = (1-p)^{1/t_p}, \; t \leqslant t_p$$

or

$$\bar{F}(t) \geqslant (1-p)^{t/t_p}, \; t \leqslant t_p. \tag{4.38}$$

Naturally, the reverse inequality is satisfied for $t > t_p$.

Thus if we know, for instance, that for $p = 0.9$, $t_p = 100$ hours, we have a lower bound for the probability that the components last more than 90 hours. Since $90 < 100$, we have $\bar{F}(90) \geqslant (0.1)^{9/10}$.

Example 8

The bound in (4.38) is sharp, in the sense that it is actually attained by the exponential distribution. For if $\bar{F}(t) = \exp(-\lambda t)$ and $\exp(-\lambda t_p) = 1-p$, then indeed $(1-p)^{t/t_p} = \exp(-\lambda t)$.

Example 9

If the continuous random variable T has p.d.f. $f(t) = 2/(t+1)^3$, $t > 0$, then $\bar{F}(t) = 1/(1+t)^2$. Hence $r(t) = f(t)/\bar{F}(t) = 2/(1+t)$ and is decreasing in t. Thus the distribution is D.F.R. and $\bar{F}(t) \leqslant (1-p)^{t/t_p}$, $t \leqslant t_p$. But $1/(1+t_p)^2 = 1-p$, or $t_p = (1/\sqrt{1-p})-1$.

A varieity of bounds is discussed by Barlow and Proschan [2]. The next problem provides another such bound, based on the mean of a distribution.

Problem 43. The random variable T has an I.F.R. distribution with mean μ. Using the inequality $E[g(T)] \leqslant g[E(T)]$ for any concave function $g(.)$, show that $\bar{F}(t) \geqslant \exp(-t/\mu)$ for $t < \mu$. ∎

REFERENCES

[1] *Applied Probability Models with Optimization Applications,* S. M. Ross, Holden-Day, Inc., 1970.

[2] *Statistical Theory of Reliability and Life Testing,* R. E. Barlow and F. Proschan. Holt, Rinehart and Winston, 1975.

[3] *Renewal Theory,* D. R. Cox. Methuen, 1962.

BRIEF SOLUTIONS AND COMMENTS ON THE PROBLEMS

Problem 1. X_i has a $\Gamma(2,\lambda)$ distribution, hence $\sum_1^n X_i$ has a $\Gamma(2n,\lambda)$ distribution

and $\sum_1^{n+1} X_i$ a $\Gamma(2n+2,\lambda)$ distribution.

$$
\begin{aligned}
F_{n+1}(t) &= \int_0^t \frac{\lambda(\lambda x)^{2n+1} \exp(-\lambda x)}{(2n+1)!}\, dx \\
&= \left[\frac{-(\lambda x)^{2n+1} \exp(-\lambda x)}{(2n+1)!} \right]_0^t + \int_0^t \frac{\lambda(\lambda x)^{2n} \exp(-\lambda x)}{2n!}\, dx \\
&= -\frac{(\lambda t)^{2n+1} \exp(-\lambda t)}{(2n+1)!} + \int_0^t \frac{\lambda(\lambda x)^{2n} \exp(-\lambda x)}{2n!}\, dx \\
&= -\frac{(\lambda t)^{2n+1} \exp(-\lambda t)}{(2n+1)!} - \frac{(\lambda t)^{2n} \exp(-\lambda t)}{2n!} + F_n(t).
\end{aligned}
$$

Problem 2. The expected cost is $c_0 + c_1 E(N_{t_0})$. If the part is *replaced* at $t_0/2$, we have two ordinary renewal processes for each interval of duration $t_0/2$ for which the expected cost is $2c_0 + 2c_1 E(N_{t_0}/2)$. The difference, which is to be < 0, is,

$$
\begin{aligned}
c_0 + \frac{c_1}{4} \{ 2[\lambda t_0 - (1-e^{-\lambda t_0})] - 2\lambda t_0 + (1-e^{-2\lambda t_0}) \} \\
= \frac{1}{4} [4c_0 - c_1(1-e^{-\lambda t_0})^2].
\end{aligned}
$$

Problem 3. From $Pr[N_t = n] = F_n(t) - F_{n+1}(t)$

$$
\begin{aligned}
E(N_t) &= \sum_1^\infty n[F_n(t) - F_{n+1}(t)] \\
&= \lim_{m \to \infty} \left[\sum_1^m nF_n(t) - \sum_1^m nF_{n+1}(t) \right] \\
&= \lim_{m \to \infty} \left[\sum_1^m nF_n(t) - \sum_2^{m+1} (n-1)F_n(t) \right] \\
&= \lim_{m \to \infty} \left[F_1(t) + \sum_2^m F_n(t) - mF_{m+1}(t) \right] \\
&= \lim_{m \to \infty} \left[\sum_1^m F_n(t) \right] - \lim_{m \to \infty} [mF_{m+1}(t)]
\end{aligned}
$$

$$= \sum_{1}^{\infty} F_n(t).$$

Problem 4

$$Pr\left[\sum_{1}^{r} X_i \leqslant t\right]$$

$$= 1 - Pr\left(\sum_{1}^{r} X_i > t\right) \leqslant 1 - Pr(t/r < X_i < t_0, \ i = 1, 2, \ldots, r)$$

$$< 1, \quad \text{since } Pr(t/r < X_i < t_0) \neq 1.$$

Divide $\sum_{1}^{jr} X_i$ into sets of r, $X_1 + X_2 + \ldots + X_r, X_{r+1} \ldots + X_{2r}$ and so on and apply equation (4.5).

Problem 5. We pick out the coefficient of x^n in $g(x) + \omega g(\omega x) + \omega^2 g(\omega^2 x)$, which is $a_n [1 + \omega.\omega^n + \omega^2(\omega^2)^n] = a_n(1 + \omega^{n+1} + \omega^{2n+2})$. If $n + 1$ is of the form $3k$, this reduces to 3, since $\omega^3 = 1$. If $n + 1$ is of the form $3k+1$, then this reduces to $a_n(1 + \omega + \omega^2) = 0$. Similarly if $n + 1$ is of the form $3k + 2$.

Problem 6

$$f^*(s) = \left(\frac{\lambda_1}{\lambda_1 + s}\right)\left(\frac{\lambda_2}{\lambda_2 + s}\right),$$

hence from equation (4.9) the Laplace transform of $E(N_t)$ is

$$\frac{\lambda_1 \lambda_2 / [(\lambda_1 + s)(\lambda_2 + s)]}{s\{1 - \lambda_1 \lambda_2 / [(\lambda_1 + s)(\lambda_2 + s)]\}} = \frac{\lambda_1 \lambda_2}{s[(\lambda_1 + \lambda_2)s + s^2]}$$

$$= \frac{\lambda_1 \lambda_2}{(\lambda_1 + \lambda_2)s^2} - \frac{\lambda_1 \lambda_2}{(\lambda_1 + \lambda_2)^2 s} + \frac{\lambda_1 \lambda_2}{(\lambda_1 + \lambda_2)^2(\lambda_1 + \lambda_2 + s)}$$

$$= \frac{\lambda_1 \lambda_2}{(\lambda_1 + \lambda_2)}\left[\frac{1}{s^2} - \frac{1}{(\lambda_1 + \lambda_2)s} + \frac{1}{(\lambda_1 + \lambda_2)(\lambda_1 + \lambda_2 + s)}\right].$$

This last expression is the Laplace transform of the displayed formula. Check $\lambda_2 = \lambda_1$ with Example 3.

Problem 7

$$f^*(s) = \frac{1}{2(s+1)} + \frac{1}{s+2} = \frac{3s+4}{2(s+1)(s+2)}.$$

The Laplace transform of $E(N_t)$ is $\dfrac{1}{s} \displaystyle\sum_1^\infty \left[\dfrac{3s+4}{2(s+1)\,(s+2)}\right]^n = \dfrac{1}{s} \cdot \dfrac{3s+4}{2s^2+3s}$

$$= \frac{4}{3s^2} + \frac{1}{9s} - \frac{2}{9(2s+3)} \ .$$

Problem 8. We give a list of part results to help the reader check a tiresome computation.

$$\int_0^t \exp(-x)\mathrm{d}x = [1-\exp(-t)]$$

$$\int_0^t x\exp(-x)\mathrm{d}x = -t\exp(-t) + \int_0^t \exp(-x)\mathrm{d}x$$

$$\int_0^t \exp[-3(t-x)/2]\,\exp(-x)\mathrm{d}x = 2\exp(-3t/2)\,[\exp(t/2)-1]$$

$$\int_0^t \exp(-2x)\mathrm{d}x = [1-\exp(-2t)]/2$$

$$\int_0^t x\exp(-2x)\mathrm{d}x = -\frac{t\exp(-2t)}{2} + \frac{1}{2}\int_0^t \exp(-2x)\mathrm{d}x$$

$$\int_0^t \exp[(-3t+3x)/2]\,\exp(-2x)\mathrm{d}x = 2\exp(-3t/2)\,[1-\exp(-t/2)] \ .$$

Problem 9

$$\int_0^\infty \left[\int_0^t H(t-x)f(x)\mathrm{d}x\right] \exp(-st)\mathrm{d}t$$

$$= \int_0^\infty \left[\int_x^\infty H(t-x)\exp(-st)\mathrm{d}t\right] f(x)\mathrm{d}x$$

$$= \int_0^\infty \left[\int_x^\infty H(t-x)\exp(-st+sx)\mathrm{d}t\right] \exp(-sx)f(x)\mathrm{d}x$$

$$= \int_0^\infty \left[\int_0^\infty H(v)\exp(-sv)\mathrm{d}v\right] f(x)\exp(-sx)\mathrm{d}x$$

$$= \int_0^\infty H^*(s)f(x)\exp(-sx)\mathrm{d}x = H^*(s)f^*(s).$$

Hence on taking the Laplace transform of both sides of equation (4.10)

$$H^*(s) = H^*(s)f^*(s) + F^*(s), \text{ or}$$

$$H^*(s)\,[1 - f^*(s)] = f^*(s)/s, \quad \text{which is equation (4.9).}$$

Problem 10

$$\int_0^t K(t-y)f(y)\mathrm{d}y = \int_0^t \left[\int_0^{t-y} k(t-y-x) \sum_1^\infty f_n(x)\mathrm{d}x + k(t-y) \right] f(y)\mathrm{d}y.$$

Changing variables, say to $v = t-y-x$, $u = y$, noting that $0 < x < t-y$ and $0 < v < t-u$, this integral becomes

$$\int_0^t \left[\int_0^{t-u} k(v) \sum_1^\infty f_n(t-v-u)\mathrm{d}v + k(t-u) \right] f(u)\mathrm{d}u.$$

Interchanging the order of integration, we have

$$\int_0^t \left[\int_0^{t-v} k(v) \sum_1^\infty f_n(t-u-v)f(u)\mathrm{d}u \right] \mathrm{d}v + \int_0^t k(t-u)f(u)\mathrm{d}u.$$

But

$$\int_0^{t-v} f_n(t-u-v)f(u)\mathrm{d}u = f_{n+1}(t-v)$$

and after adjusting for the term $n = 1$, we have $K(t) - k(t)$.

Problem 11

$$\int_0^\infty \left[\int_0^t F_{n-1}(t-x_1)f(x_1)\mathrm{d}x_1 \right] \exp(-st)\mathrm{d}t$$

$$= \int_0^\infty \left[\int_{x_1}^\infty F_{n-1}(t-x_1)\exp(-st)\mathrm{d}t \right] f(x_1)\mathrm{d}x_1$$

$$= \int_0^\infty \left[\int_{x_1}^\infty F_{n-1}(t-x_1)\exp(-st+sx_1)\mathrm{d}t \right] f(x_1)\exp(-sx_1)\mathrm{d}x_1$$

$$= F_{n-1}^*(s)f^*(s).$$

Hence, from equation (4.8), this is

$$\frac{1}{s}\,[f^*(s)]^{n-1} f^*(s) = F_n^*(s).$$

Problem 12. Start with

$$Pr(\sum_1^n X_i \leqslant t | X_1 = x_1) = Pr(\sum_2^n X_i \leqslant t - x_1 | X_1 = x_1)$$

$$= Pr(\sum_{2}^{n} X_i \leqslant t - x_1).$$

Then $Pr(\sum_{1}^{n} X_i \leqslant t) = \int_{0}^{t} Pr(\sum_{1}^{n} X_i \leqslant t | X_1 = x_1) g(x_1) dx_1.$

This gives the first equation. The alternative form of the result follows by changing the variable to $t - x_1$. The probability density function is found by differentiating the first form with respect to t giving, since $F_{n-1}(0) = 0$,

$$\int_{0}^{t} f_{n-1}(t - x_1) g(x_1) dx_1.$$

Problem 13

(i) $\displaystyle \int_{0}^{t} H(t-x) \sum_{1}^{\infty} f_n(x) dx = \int_{0}^{t} \sum_{1}^{\infty} F_m(t-x) \sum_{1}^{\infty} f_n(x) dx$

$$= \sum_{1}^{\infty} \sum_{1}^{\infty} \int_{0}^{t} F_m(t-x) f_n(x) dx$$

$$= \sum_{1}^{\infty} \sum_{1}^{\infty} F_{m+n}(t).$$

Not put $m + n = k$, $n = \ell$, noting that $\ell \geqslant 1$, $k \geqslant 2$. We have $1 \leqslant \ell \leqslant k - 1$.

$$\sum_{k=2}^{\infty} \sum_{\ell=1}^{k-1} F_k(t) = \sum_{k=2}^{\infty} (k-1) F_k(t).$$

(ii) $V(N_t) = E(N_t^2) - [E(N_t)]^2.$

$$E(N_t^2) = \sum_{1}^{\infty} n^2 Pr(N_t = n) = \sum_{1}^{\infty} n^2 [F_n(t) - F_{n+1}(t)].$$

Now $\displaystyle \sum_{1}^{m} n^2 F_n(t) - \sum_{1}^{m} n^2 F_{n+1}(t) = \sum_{1}^{m} n^2 F_n(t) - \sum_{2}^{m+1} (n-1)^2 F_n(t).$

$$= F_1(t) + \sum_{2}^{m} (2n-1) F_n(t) - m^2 F_{m+1}(t)$$

$$= \sum_{1}^{m} F_n(t) + 2 \sum_{2}^{m} (n-1) F_n(t) - m^2 F_{m+1}(t).$$

Now take limit as $m \to \infty$ and use the result in (i).

Problem 14

$$\frac{t}{\mu} = F(t) + \frac{t}{\mu} \int_{0}^{t} f(x) dx - \frac{1}{\mu} \int_{0}^{t} x f(x) dx.$$

Differentiating with respect to t,

$$\frac{1}{\mu} = f(t) + \frac{1}{\mu} \int_0^t f(x)dx + \frac{t}{\mu} f(t) - \frac{1}{\mu} tf(t).$$

$$\frac{1}{\mu} = F'(t) + \frac{1}{\mu} F(t) \Rightarrow \frac{d}{dt} [\exp(t/\mu)F(t)] = \exp(t/\mu)/\mu.$$

Problem 15. For Problem 6, directly from the properties of the exponential distribution, the inter-renewal time has mean $\frac{1}{\lambda_1} + \frac{1}{\lambda_2}$ and variance $\frac{1}{\lambda_1{}^2} + \frac{1}{\lambda_2{}^2}$.

For Problem 7, $\mu = E(X) = \frac{1}{2} + \frac{1}{2}\left(\frac{1}{2}\right) = \frac{3}{4}$. $E(X^2) = \frac{1}{2}(2) + \frac{1}{2}\left(\frac{1}{2}\right) = \frac{5}{4}$.

$$\sigma^2 = \frac{5}{4} - \frac{9}{16} = \frac{11}{16}.$$

Problem 16. $G^*(s) = H^*(s) - 1/s\mu s^2$. Hence, using (4.18),

$$sG^*(s) = \frac{f^*(s)}{1 - f^*(s)} - \frac{1}{s\mu} = \frac{s\mu f^*(s) - 1 + f^*(s)}{s\mu[1 - f^*(s)]}.$$

since $f^*(0) = 1$, $\lim_{s \to 0} [sg^*(s)]$ needs some care and L'Hôpital's rule must be applied twice. In the course of evaluating the limit the reader will require (from Appendix 2 or otherwise)

$$\lim_{s \to 0} \left[\frac{d}{ds} f^*(s)\right] = -\mu, \lim_{s \to 0} \left[\frac{d^2}{ds^2} \{f^*(s)\}\right] = \mu^2 + \sigma^2.$$

Hence we obtain $\lim_{t \to 0} [G(t)]$. Unfortunately this is not enough to give us $\lim_{t \to \infty} H(t)$.

Problem 17. The Laplace transform is (see equation (4.7))

$$E[\exp(-\Sigma X_i s)] = E[\exp(-X_1 s)] \prod_2^n E[\exp(-X_i s)]$$

$$= g^*(s) [f^*(s)]^{n-1}.$$

$$H_m^*(s) = \frac{1}{s} \cdot \sum_1^\infty g^*(s) [f^*(s)]^{n-1} = \frac{1}{s} \cdot \frac{g^*(s)}{[1 - f^*(s)]}.$$

Problem 18

$$\int_0^t H_m(t-x)f(x)dx = \int_0^t \sum_1^\infty F_n^{(m)} (t-x)f(x)dx$$

$$= \sum_1^\infty \int_0^t F_n^{(m)}\,(t-x)f(x)\mathrm{d}x = \sum_1^\infty F_{n+1}^{(m)}(t) = H_\mathrm{m}(t) - F_1^{(m)}(t)$$

$$= H_\mathrm{m}(t) - G(t).$$

From Problem 17, $H_\mathrm{m}^*(s) = \dfrac{1}{s} \cdot \dfrac{g^*(s)}{[1-f^*(s)]}$, where

$g^*(s) = \lambda_1/(\lambda_1+s)$, $f^*(s) = \lambda_1\lambda_2/[(\lambda_1+s)\,(\lambda_2+s)]$. The Laplace transform of the displayed formula is

$$\frac{\lambda_1\lambda_2}{(\lambda_1+\lambda_2)s^2} + \frac{\lambda_1^2}{(\lambda_1+\lambda_2)^2}\left(\frac{1}{s} - \frac{1}{\lambda_1+\lambda_2+s}\right).$$

Problem 19

$$\frac{t}{\mu} = G(t) + \int_0^t \left(\frac{t-x}{\mu}\right) f(x)\mathrm{d}x.$$

Integrate by parts

$$\int_0^t \left(\frac{t-x}{\mu}\right) f(x)\mathrm{d}x = \left[\frac{t-x}{\mu} F(x)\right]_0^t + \frac{1}{\mu}\int_0^t F(x)\mathrm{d}x.$$

Hence $G(t) = \dfrac{1}{\mu}\displaystyle\int_0^t [1-F(x)]\,\mathrm{d}x$. Finally differentiate both sides with respect to t.

Problem 20

$$\int_0^t \left[\sum_1^\infty f_n(t-x)\right] f(x)\mathrm{d}x = \sum_1^\infty \left[\int_0^t f_n(t-x)f(x)\mathrm{d}x\right]$$

$$= \sum_1^\infty f_{n+1}(t) = h_\mathrm{o}(t) - f(t).$$

$$h_\mathrm{o}^*(s) = \sum_1^\infty f_n^*(s) = \sum_1^\infty [f^*(s)]^n = f^*(s)/[1-f^*(s)].$$

Problem 21. For the $\Gamma(4,\lambda)$ distribution, $f^*(s) = [\lambda/(\lambda+s)]^4$. From Problem 20,
$$h_\mathrm{o}^*(s) = \frac{[\lambda/(\lambda+s)]^4}{1-[\lambda/(\lambda+s)]^4}$$

$$= \frac{\lambda^4}{(\lambda+s)^4-\lambda^4} = \frac{\lambda^4}{[(\lambda+s)^2-\lambda^2]\,[(\lambda+s)^2+\lambda^2]}$$

$$= \frac{\lambda^2}{2}\left[\frac{1}{(\lambda+s)^2-\lambda^2} - \frac{1}{(\lambda+s)^2+\lambda^2}\right].$$

Venturing further into partial fractions,

$$\frac{1}{(\lambda+s)^2 - \lambda^2} = \frac{1}{2\lambda}\left(\frac{1}{s} - \frac{1}{s+2\lambda}\right).$$

$$\frac{1}{(\lambda+s)^2 + \lambda^2} = \frac{1}{2\omega\lambda}\left[\frac{1}{(\lambda+s)-\omega\lambda} - \frac{1}{(\lambda+s)+\omega\lambda}\right]$$

where $\omega = \sqrt{-1}$. The functions which have Laplace transforms $\frac{1}{s}$, $\frac{1}{2\lambda+s}$, $\frac{1}{\lambda(1-\omega)+s}$, $\frac{1}{\lambda(1+\omega)+s}$, are 1, $\exp(-2\lambda t)$, $\exp[-\lambda(1-\omega)t]$, $\exp(-\lambda(1+\omega)t]$. Bear in mind that $\exp(-\omega\lambda t) = \cos(\lambda t) - \omega\sin(\lambda t)$. Alternatively take the Laplace transform of $h_0(t)$, consulting Appendix 2 when necessary.

Problem 22

$$\int_0^t \sum_1^\infty f_n^{(m)}(t-x)f(x)\mathrm{d}x = \sum_1^\infty \int_0^t f_n^{(m)}(t-x)f(x)\mathrm{d}x$$

$$= \sum_1^\infty f_{n+1}^{(m)}(t) = h_m(t) - f_1^{(m)}(t) = h_m(t) - g(t).$$

Problem 23

$$\int_0^t h_m(t-u)\,[1-F(u)]\,\mathrm{d}u = \left[-H_m(t-u)\right]_0^t - \sum_1^\infty \int_0^t f_n^{(m)}(t-u)F(u)\mathrm{d}u$$

$$= H_m(t) - \sum_1^\infty F_{n+1}^{(m)}(t) = H_m(t) - H_m(t) + F_1^{(m)}(t)$$

$$= G(t).$$

Problem 24

$$h_0(t) = \lambda, \text{ and } 1-F(u) = \exp(-\lambda u),$$

$$\int_0^t u.\,\lambda\exp(-\lambda u)\mathrm{d}u = \left[-u\exp(-\lambda u)\right]_0^t + \int_0^t \exp(-\lambda u)\mathrm{d}u$$

$$= -t\exp(-\lambda t) + [1-\exp(-\lambda t)]/\lambda.$$

Problem 25. For the discrete probability at t, see solution to Problem 19. The continuous part is obtained immediately from equation (4.30) with $h_m(t-u) = 1/\mu$.

Problem 26

$$h_m(y) = \lambda.$$

$$\int_0^t \lambda^2 \exp[-\lambda(t+v-y)]\, dy = \lambda\,[\exp(-\lambda v) - \exp(-\lambda t - \lambda v)] \text{ and } g(v+t) =$$

$$\lambda \exp(-\lambda t - \lambda v).$$

Problem 27. $f(x)$ is the p.d.f. of a $\Gamma(2,\lambda)$ distribution and $f^*(s) = [\lambda/(\lambda+s)]^2$.

$$h_o^*(s) = \frac{[\lambda/(\lambda+s)]^2}{1 - [\lambda/(\lambda+s)]^2} = \frac{\lambda^2}{(\lambda+s)^2 - \lambda^2} = \frac{1}{2}\left(\frac{\lambda}{s} - \frac{\lambda}{2\lambda+s}\right).$$

Hence $h_o(y) = \dfrac{\lambda}{2} - \dfrac{\lambda}{2}\exp(-2\lambda y).$

$$\psi_o(v) = \lambda^2\,(v+t)\exp[-\lambda(v+t)] + \int_0^t \frac{1}{2}\lambda\,[1 - \exp(-2\lambda y)]$$

$$\lambda^2\,(t+v-y)\exp[-\lambda(t+v-y)]\, dy.$$

Problem 28. If the first part has not failed by time t, then the backward recurrence time is t and for the forward recurrence time to exceed t, its life must be at least $2t$. This happens with probability $1-F(2t)$. Otherwise, there are $n(\geqslant 1)$ renewals by time t, the nth renewal takes place at time $x(<t)$, and the part then installed fails after a further time $y(>t-x)$. We require $t-x < y-(t-x)$ or $y > 2t-2x$. This event has probability

$$\int_0^t [1 - F(2t-2x)]\, f_n(x)dx.$$

Now sum over n.

Problem 29. From the equilibrium process, $h_e(y) = 1/\mu$. Hence, from (4.31),

$$\psi_e(v) = g(v+t) + \frac{1}{\mu}\left[-F(t+v-y)\right]_0^t$$

$$= g(v+t) + \frac{1}{\mu}\,[F(t+v) - F(v)].$$

But, from 4.25, $g(t) = \dfrac{1}{\mu}\,[1 - F(t)].$

Problem 30. For the $\Gamma(2,\lambda)$ distribution,

$$\bar{F}(t) = \int_t^\infty \lambda(\lambda y)\exp(-\lambda y)dy = \left[-\lambda y\exp(-\lambda y)\right]_t^\infty + \int_t^\infty \lambda\exp(-\lambda y)dy$$

$$= (1+\lambda t)\exp(-\lambda t).$$

Problem 31. Consider $\bar{F}(t_1 - t_0)\,\bar{F}(t_0) - \bar{F}(t_1)$, where

$$\bar{F}(t_1 - t_0) = 1 - p)\exp(-\lambda_1 t_1 + \lambda_1 t_0) + p\exp(-\lambda_2 t_1 + \lambda_2 t_0).$$

After collecting terms, we obtain

$$p(1-p)\left[e^{-\lambda_1 t_0} - e^{-\lambda_2 t_0}\right]\left[e^{-\lambda_2(t_1 - t_0)} - e^{-\lambda_1(t_1 - t_0)}\right] < 0$$

whether $\lambda_1 < \lambda_2$ or $\lambda_2 < \lambda_1$.

Problem 32. For the Weibull distribution, $\bar{F}(x) = \exp[-(\lambda x)^\alpha]$.
Hence $\bar{F}(t_1 - t_0)\bar{F}(t_0) - \bar{F}(t_1)$

$$= \exp[-\lambda^\alpha(t_1 - t_0)^\alpha - \lambda^\alpha t_0^\alpha] - \exp(-\lambda^\alpha t_1{}^\alpha)$$

$$= \exp(-\lambda^\alpha t_1{}^\alpha)\,[\exp\{\lambda^\alpha(t_1{}^\alpha - t_0{}^\alpha - (t_1 - t_0)^\alpha)\} - 1].$$

Evidently the sign of this expression depends on the sign of

$$t_1{}^\alpha - t_0{}^\alpha - (t_1 - t_0)^\alpha = t_1{}^\alpha\,[1 - x^\alpha - (1-x)^\alpha],$$

where $0 < x = t_0/t_1 < 1$. $x^\alpha + (1-x)^\alpha$ has a turning value at $x = \frac{1}{2}$. If $\alpha > 1$, $x^\alpha + (1-x)^\alpha$ has a minimum and lies between 1 and $2(\frac{1}{2})^\alpha = (\frac{1}{2})^{\alpha - 1} < 1$. If $\alpha < 1$, $x^\alpha + (1-x)^\alpha$ has a maximum and lies between 1 and $2^{1-\alpha} > 1$. Hence if $\alpha > 1$, the distribution is N.B.U. and if $\alpha < 1$ the distribution is N.W.U.

Problem 33. The probability that the first failure has not appeared by time t where $kt_a < t < (k+1)t_a$ is $[\bar{F}(t_a)]^k\,\bar{F}(t - kt_a)$. Hence the expected time to the first failure is

$$\sum_{k=0}^{\infty}\int_{kt_a}^{(k+1)t_a}[\bar{F}(t_a)]^k\,\bar{F}(t - kt_a)\mathrm{d}t$$

$$= \sum_{k=0}^{\infty}\{\bar{F}(t_a)\}^k\int_0^{t_a}\bar{F}(y)\mathrm{d}y$$

$$= \frac{1}{F(t_a)}\int_0^{t_a}\bar{F}(y)\mathrm{d}y.$$

Problem 34. From (4.32), $\bar{F}(t_1 - t_0)\bar{F}(t_0) > \bar{F}(t_1)$. If $t_1 = (k+1)t_a$, $t_0 = t_a$, $\bar{F}(kt_a)\bar{F}(t_a) > \bar{F}[(k+1)t_a]$. Hence $[\bar{F}(t_a)]^m = [\bar{F}(t_a)]^{m-2}\,\bar{F}(t_a)\bar{F}(t_a) > [\bar{F}(t_a)]^{m-2}\,\bar{F}(2t_a)$ and the result follows by repeated application.

We have $kt_a < t < (k+1)t_a$ and $\ell m t_a < t < (\ell+1)m t_a$. Hence $[\bar{F}(t_a)]^k = [\bar{F}(t_a)]^{m\ell + k - m\ell} > [\bar{F}(mt_a)]^\ell\,\bar{F}[(k - m\ell)t_a]$.
Therefore, $[\bar{F}(t_a)]^k\,\bar{F}(t - kt_a) > [\bar{F}(mt_a)]^\ell\,\bar{F}(t - kt_a)\,\bar{F}[(k - m\ell)t_a]$

$$> [\bar{F}(mt_a)]^\ell\,\bar{F}(t - m\ell t_a).$$

(Since k, ℓ, m are integers, it readily follows that $(\ell+1)m \geqslant k+1$.)

Problem 35

$$f(t)/\bar{F}(t) = \left[\int_t^\infty \left(\frac{x}{t}\right)^{\alpha-1} \exp\{-\lambda(x-t)\}\, dx \right]^{-1}$$

$$= \left[\int_0^\infty (1+y/t)^{\alpha-1} \exp(-\lambda y) dy \right]^{-1}.$$

Since $(1 + y/t)^{\alpha-1}$ decreases or increases with t according as $\alpha > 1$ or $\alpha < 1$, the result follows.

Problem 36

$$r(t) = f(t)/[\bar{F}(t)]$$

$$= -\frac{d}{dt}[\log \bar{F}(t)].$$

Hence $\quad -\int_0^t r(y) dy = \log[\bar{F}(t)],$

$$\bar{F}(t) = \exp\left[-\int_0^t r(y) dy\right],$$

Hence $\quad \dfrac{\bar{F}(t+x)}{\bar{F}(t)} = \exp\left[-\int_t^{t+x} r(y) dy\right],$

$$= \exp\left[-\int_0^x r(u+t) du\right]$$

But $r(u+t)$ is increasing in t, hence $\bar{F}(t+x)/\bar{F}(t)$ is decreasing in t.

Problem 37. From (4.37)

$$\mu > \int_t^\infty \bar{F}(x) dx / \bar{F}(t) = \left[\mu - \int_0^t \bar{F}(x) dx\right]/\bar{F}(t).$$

Hence result. We have

$$\frac{t}{\mu} = H_e(t) = \sum_1^\infty F_n^{(e)}(t) = \sum_1^\infty \int_0^t G(y) f_{n-1}(t-y) dy$$

where, via Problem 19, $G(y) = \dfrac{1}{\mu} \int_0^y \bar{F}(x) dx \geq F(y)$ from the earlier result.

Hence $\dfrac{t}{\mu} \geqslant \sum_1^\infty \int_0^t F(y) f_{n-1}(t-y) dy = H_0(t).$

Problem 38. For the $\Gamma(2,\lambda)$ distribution, $H_0(t) = [2\lambda t - \{1-\exp(-2\lambda t)\}]/4$.

For scheme I, $c_s \left[\dfrac{\lambda}{2} - \dfrac{\{1-\exp(-2\lambda k t_0)\}}{4kt_0} \right]$.

For scheme II, $\dfrac{c_p}{t_0} + c_s \left[\dfrac{\lambda}{2} - \dfrac{\{1-\exp(-2\lambda t_0)\}}{4t_0} \right]$.

Problem 39

Scheme 1 $c_s \left[\dfrac{kt_0}{\mu} + \dfrac{\sigma^2-\mu^2}{2\mu^2} + o(1) \right] /kt_0$,

Scheme II $\dfrac{c_p}{t_0} + c_s \left[\dfrac{t_0}{\mu} + \dfrac{\sigma^2-\mu^2}{2\mu^2} + o(1) \right] /t_0$.

$\Rightarrow$ Cost scheme I minus cost Scheme II is, for large t_0 approximately

$$c_s \left(\dfrac{\sigma^2-\mu^2}{2\mu^2} \right) \left(\dfrac{1}{kt_0} - \dfrac{1}{t_0} \right) - \dfrac{c_p}{t_0} \ .$$

Problem 40. The expected number of age-replacements before the first failure is $\bar{F}(t_a)/F(t_a)$. The expected time to the first failure is $\int_0^{t_a} \bar{F}(y)dy/F(t_a) = \mu^*$ (see Problem 33). Hence the ratio of the expected cost up to the first failure to the expected time to the first failure is

$$\{[\bar{F}(t_a)/F(t_a)] c_a + c_s\} /\mu^*.$$

After substituting for μ^* we obtain the result.

Problem 41

$$F(t_a) \to 1, \bar{F}(t_a) \to 0, \int_0^{t_a} \bar{F}(x)dx \to \mu,$$

hence, expected cost per unit time $\to c_s/\mu$. Since $\int_0^{t_a} \bar{F}(x)dx \leqslant t_a$, the expected cost per unit time $\geqslant c_a/t_a$. If the components are I.F.R. and there is a finite minimum, it is not worth making age replacements if $c_s/\mu \leqslant c_a/t_a$.

Problem 42. Differentiate with respect to t_0. If a (finite) minimum exists, it satisfies $t_0 h(t_0) - H(t_0) = c_p/c_s$. For the $\Gamma(2,\lambda)$ distribution, $H(t) = [2\lambda t - 1 + \exp(-2\lambda t)]/4$ and $h(t) = H'(t)$. After substituting, $(1 + 2\lambda t_0) \exp(-2\lambda t_0) = 1 - 4c_p/c_s$. If $4c_p > c_s$, no block replacements should be made.

Problem 43. Since T has an I.F.R. distribution, $\log \bar{F}(t)$ is concave. Hence $E[\log \bar{F}(T)] \leqslant \log \bar{F}[E(T)] = \log \bar{F}(\mu)$.

$$E[\log \bar{F}(T)] = \int_0^\infty \log [\bar{F}(t)]\, f(t)\,\mathrm{d}t. \text{ Substituting } \bar{F}(t) = x,$$

$$E[\log \bar{F}(T)] = \int_0^1 \log x.\,\mathrm{d}x = \left[x \log (x)\right]_0^1 - \int_0^1 1\,\mathrm{d}x = -1.$$

We conclude that $\bar{F}(\mu) \geqslant \exp(-1)$.

But if $t < \mu$, $[\bar{F}(\mu)]^{1/\mu} \leqslant [\bar{F}(t)]^{1/t}$, hence result.

CHAPTER 5

Reliability of Systems

5.1 INTRODUCTION

Under the topic 'Renewal Theory' we have mostly been concerned with a few components and their replacement. A system may, however, depend on the functioning of many components and, moreover, do so in a rather complicated way. We shall simplify matters by assuming that different components function independently of each other. This is indeed a sweeping assumption since, in a mechanical system, the failure of an individual part is likely to throw an extra strain on the remainder. We are not now concerned with replacement or repairs. Hence, to say that a system, or a component, is functioning at time t also implies that it has done so throughout the interval $(0,t)$. Similarly, if it is not functioning at t, then it has not done so for some duration before t. Suppose in particular that the time to failure of the ith component has cumulative distribution function $F_i(t)$. Then the probability that it is still functioning at t is $1-F_i(t) = p_i(t)$, which we call the **reliability** of the ith component. The reliability of the whole system is the probability that *it* is still functioning at time t. The structure of the system tells us which combinations of working parts enable the system to function and which do not. Thus if there are two independent components, with reliabilities $p_1(t)$, $p_2(t)$, there are only two possible structures.

(i) The system functions if and only if both components function. The reliability of the system is thus $p_1(t)p_2(t)$.
(ii) The system functions if and only if at least one of the components is functioning. Since the probability that both components have failed is $(1-p_1(t)][1-p_2(t)]$, the reliability of this sytem is $1-[1-p_1(t)][1-p_2(t)]$.

5.2 NETWORKS

There is an illuminating analogy between such systems and electrical networks. In the network, we allocate a switch to each component in the system. If a component is functioning, we say the corresponding switch is closed and permits current to flow while if the component is not functioning the switch is open and current cannot pass. The system will function as a whole if there is a route through the network through which current can flow. Thus the network corresponding to (i) is shown in Fig. 5.1. Current flows if and only if both switches are closed and it is appropriate to say that the components are in **series**.

Figure 5.1

Corresponding to (ii) the network is as shown in Fig. 5.2. and current flows if and only if at least one switch is closed, and we say that the components are in **parallel**. Many systems may be described by suitable combinations of components in series and/or in parallel. So long as we are considering the functioning of a system at a fixed time, we may suppress the 't' in the notation for the various reliabilities.

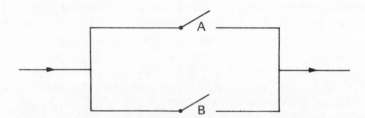

Figure 5.2

Example 1
A system consists of four components A,B,C,D each of which has reliability p independently of the state of the others. The system functions if and only if at least one of A,B functions *and* at least one of C,D functions. One way of computing the reliability of the system is to tackle directly the probability of the event which guarantees that the system is working. We consider then

$$Pr\ [(A\ or\ B)\ and\ (C\ or\ D)] = Pr\ [(A\ or\ B)]\ Pr\ [(C\ or\ D)]$$

$$= [1-Pr\ (not\ A\ and\ not\ B)]\ [1-Pr\ (not\ C\ and\ not\ D)]$$

$$= [1-Pr\ (not\ A)\ Pr\ (not\ B)]\ [1-Pr\ (not\ C)\ Pr\ (not\ D)]$$

$$= [1-(1-p)(1-p)]\ [1-(1-p)(1-p)]$$

$$= [1-(1-p)^2]^2. \tag{5.1}$$

On the other hand, the network representation invites us to regard A,B as a parallel pair in series with another parallel pair C,D (see Fig. 5.3). Current will flow unless (i) both switches A and B are open, or (ii) both C and D are open. Thus the probability that current does not flow through the circuit is,

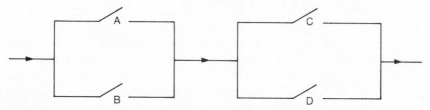

Figure 5.3

$$(1-p)^2 + (1-p)^2 - (1-p)^2 (1-p)^2,$$

and the probability that it does flow is

$$1 - [2(1-p)^2 - (1-p)^4] = [1-(1-p)^2]^2.$$

The attractions of a network representation emerge even more forcibly for more complicated systems.

Problem 1. A system consists of four independent components A,B,C,D, each with reliability p. The system functions if and only if either both A and B or both C and D function. Draw a network diagram and show that the reliability of the system is $p^2(2-p^2)$. ■

Problem 2. Show that the reliability of the system in Example 1 exceeds the individual reliability of its components if $p > 0.38$. For what values of p is the system more reliable then the arrangement of components given in Problem 1? ■

Problem 3. We reconsider Example 1 to illustrate the complications that may arise if the starting point is less advantageous. The following pairs of switches, (A,C), (A,D), (B,C), (B,D), will, when both switches are closed, ensure that the system functions. Each of the pairs has reliability p^2. Using the method of inclusion and exclusion, show that the probability that at least one of these pairs is functioning is $p^4 - 4p^3 + 4p^2 = [1-(1-p)^2]^2$. (Hint: bear in mind that not all pairs are independent of each other since they may have components in common.) ■

5.3 STRUCTURE FUNCTIONS

We now break off this somewhat ad hoc discussion of system reliability and construct a more systematic framework. Suppose the system contains n

components and to the ith of these we assign a variable, x_i, which takes the value $+1$ if that component is functioning and is zero otherwise. For each set of values $x_1, x_2, \ldots, x_n$ the design of the system determines whether it functions or not. The **structure function**, $\phi(x_1, x_2, \ldots, x_n)$, assumes the values $+1$ when the system functions and is zero otherwise.

Example 2

For Example 1, let x_1, x_2, x_2, x_4 be assigned to components A,B,C,D respectively. Since each x_i can assume two values, there are $2^4 = 16$ distinguishable sets (x_1, x_2, x_3, x_4). Now the design of the system is such that it functions so long as at least one of A,B *and* at least one of C,D functions. That is to say, the structure function takes the value $+1$ if at least one of (x_1, x_2) is $+1$ and at least one of (x_3, x_4) is $+1$. It is easily checked that the following nine sets satisfy $\phi(x_1, x_2, x_3, x_4) = 1$,

x_1	1	0	1	1	1	0	1	0	1
x_2	1	1	0	1	1	1	0	1	0
x_3	1	1	1	0	1	0	1	1	0
x_4	1	1	1	1	0	1	0	0	1 .

As a problem, it should be verified that the remaining seven sets imply that this system does not function.

5.4 FORMULAE FOR STRUCTURE FUNCTIONS

Listing the values of a function, as illustrated, scarcely encourages manipulation. What we really need is an explicit algebraic function of x_1, x_2, x_3, x_4 which assumes the value $+1$ if and only if the system is working and is zero otherwise. The starting point should surely be components in series. The obvious candidate for the structure function for n components in series is

$$\phi(x_1, x_2, \ldots, x_n) = \prod_1^n x_i, \tag{5.2}$$

for, since each x_i takes the values 0,1, the product takes the value 1 if and only if all the x_i are 1, and is zero otherwise. From (5.2) we need only a few steps to find the structure function of n units in parallel. We require

$$\phi(x_1, x_2, \ldots, x_n) = 1 \iff \text{at least one } x_i = 1,$$
or
$$\phi(x_1, x_2, \ldots, x_n) = 0 \iff \text{all } x_i = 0,$$
or
$$1 - \phi(x_1, x_2, \ldots, x_n) = 1 \iff \text{all } x_i = 0,$$
or
$$1 - \phi(x_1, x_2, \ldots, x_n) = 1 \iff \text{all } 1 - x_i = 1.$$

But then, from (5.2),

$$1-\phi(x_1,x_2,\ldots,x_n) = \prod_1^n (1-x_i),$$

or
$$\phi(x_1,x_2,\ldots,x_n) = 1- \prod_1^n (1-x_i). \tag{5.3}$$

It is now easy to check in (5.3) that this ϕ does indeed take the value $+1$ if and only if at least one $x_i = 1$.

Problem 4. Since the variables x_i take only the values $0,1$, show that for n components in series $\phi(x_1,x_2,\ldots,x_n) = \min(x_1,x_2,\ldots,x_n)$, and for n components in parallel $\phi(x_1,x_2,\ldots,x_n) = \max(x_1,x_2,\ldots,x_n)$. ∎

Problem 5. T_1,T_2 are independent random variables. Show that

$$Pr\,[\max(T_1,T_2) \geqslant t] = 1-[1-Pr(T_1 \geqslant t)]\,[1-Pr(T_2 \geqslant t)]\,.$$

Compare with the result in equation (5.3). State an analogous result for the $Pr\,[\min(T_1,T_2) \geqslant t]$ and compare with (5.2). ∎

If system I has n components and structure function $\phi_I(x_1,x_2,\ldots,x_n)$ while system II has m components with structure function $\phi_{II}(y_1,y_2,\ldots,y_m)$ and if I and II are joined in *series*, then the structure function for the resultant system is

$$\phi_I(x_1,x_2,\ldots,x_n)\,\phi_{II}(y_1,y_2,\ldots,y_m). \tag{5.4}$$

If I, II are joined in *parallel*, then the structure function for the resultant system is

$$1-[1-\phi_I(x_1,x_2,\ldots,x_n)]\,[1-\phi_{II}(y_1,y_2,\ldots,y_m)]. \tag{5.5}$$

Both (5.4) and (5.5) follow almost immediately from the considerations advanced in support of equations (5.2) and (5.3). The structure function for more complicated systems can often be found by splitting the system into subsets of components in series or in parallel.

Example 3

The system in Example 1, with circuit diagram as shown in Fig. 5.4 clearly consists of two units in parallel, with structure function $1-(1-x_1)(1-x_2)$, in series with another two units in parallel, with structure function $1-(1-x_3)(1-x_4)$. Hence the structure function for the complete system is

$$\phi(x_1,x_2,x_3,x_4) = [1-(1-x_1)(1-x_2)]\,[1-(1-x_3)(1-x_4)].$$

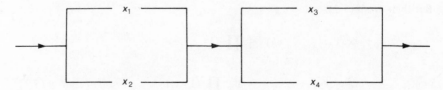

Figure 5.4

Problem 6. Show that the structure function for the system in Problem 2 is $1-(1-x_1 x_2)(1-x_3 x_4)$. ∎

The k out of n system

A system is said to be **k out of n** when it functions if and only if at least k out of its n components function. Its structure function is by no means obvious, but from its definition

$$\phi(x_1,x_2,\ldots,x_n) = 1, \quad \sum_1^n x_i \geqslant k$$

$$\phi(x_1,x_2,\ldots,x_n) = 0, \quad \sum_1^n x_i < k. \tag{5.6}$$

In this context, a series system is also n out of n and a parallel system is 1 out of n.

Example 4

To find the structure function of a 2 out of 3 system we identify all minimal sets of components which can guarantee functioning of the system.

Thus if both members of any of the pairs (x_1,x_2), (x_2,x_3), (x_3,x_1) are equal to 1, then the system functions. Variables within a pair are to behave as though in series and $x_1 x_2$, $x_2 x_3$, $x_1 x_3$ are assigned to these sub-systems. However, the system will function if at least one such sub-system functions and the three sub-systems may be thought of as in parallel. Hence the structure function for the two out of three system is

$$\phi(x_1,x_2,x_3) = 1-(1-x_1 x_2)(1-x_2 x_3)(1-x_3 x_1).$$

After multiplying out and bearing in mind that $x_i^2 = x_i$, we arrive at

$$\phi(x_1,x_2,x_3) = x_1 x_2 + x_1 x_3 + x_2 x_3 - 2x_1 x_2 x_3.$$

A circuit diagram looks like the one shown in Fig. 5.5 where all switches marked with the same variable are in the same state.

Figure 5.5

Problem 7. Show for Example 4 that the structure function can be written

$$x_1 x_2 (1-x_3) + x_1 x_3 (1-x_2) + x_2 x_3 (1-x_1) + x_1 x_2 x_3.$$

Can you give an interpretation to the various terms? ∎

Decomposition

Another tool for finding structure functions is provided by a 'stepping down' technique. We have the following identity

$$\phi(x_1,x_2,\ldots,x_n) \equiv x_1 \phi(1,x_2,\ldots,x_n) + (1-x_1)\phi(0,x_2,\ldots,x_n). \quad (5.7)$$

Equation (5.7) is easily checked by substituting $x_1 = 0,1$ in both sides. In a sense we have expanded the function about x_1. We can repeat the process in terms of x_2 in both of $\phi(1,x_2,\ldots,x_n)$ and $\phi(0,x_2,\ldots,x_n)$. At each stage the number of undeclared variables is reduced by one and eventually the stage is reached where it is possible to identify the value of all terms in the decomposition from the design of the system.

Example 5

To find the structure function of the 2 out of 3 system.

$$\begin{aligned}
\phi(x_1,x_2,x_3) &= x_1 \phi(1,x_2,x_3) + (1-x_1)\, \phi\, (0,x_2,x_3) \\
&= x_1 [x_2 \phi(1,1,x_3) + (1-x_2)\, \phi\, (1,0,x_3)] \\
&\quad + (1-x_1)\, [x_2 \phi(0,1,x_3) + (1-x_2)\, \phi\, (0,0,x_3)].
\end{aligned}$$

Now $\phi(1,1,x_3) = 1$, since two components are functioning, and $\phi(0,0,x_3) = 0$, since two components are not functioning, *whatever* the value of x_3. Similarly,

$$\phi(1,0,x_3) = x_3 \phi(1,0,1) + (1-x_3)\, \phi\, (1,0,0) = x_3$$

and

$$\phi(0,1,x_3) = x_3 \phi(0,1,1) + (1-x_3)\, \phi\, (0,1,0) = x_3.$$

Finally

$$\begin{aligned}
\phi(x_1,x_2,x_3) &= x_1 [x_2 + (1-x_2)x_3] + (1-x_1) x_2 x_3 \\
&= x_1 x_2 + x_1 x_3 + x_2 x_3 - 2 x_1 x_2 x_3.
\end{aligned}$$

The first line may be read 'either x_1 and at least one of x_2,x_3 or not x_1 and both of x_2,x_3.

Consistency of Structures

It is convenient at this point to remark on two minor but general points about structure functions. It would be disconcerting if a system performed worse when a particular component functioned than when this same component did not function. Thus we require a structure function to be **monotone**. That is to say, if $x_i \leqslant y_i$ $(i = 1, 2, \ldots, n)$ then

$$\phi(x_1, x_2, \ldots, x_n) \leqslant \phi(y_1, y_2, \ldots, y_n).$$

Moreover, it serves little purpose to include in the structure function a variable corresponding to a component which plays no role in the functioning of the system. More precisely, the $i$the component is said to be **irrelevant** if

$$\phi(x_1, x_2, \ldots, x_{i-1}, 0, x_{i+1}, \ldots, x_n) = \phi(x_1, x_2, \ldots, x_{i-1}, 1, x_{i+1}, \ldots, x_n)$$

for every other set of states for the remaining components.

Example 6

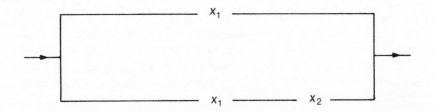

Figure 5.6

In the circuit shown in Fig. 5.6, the two switches marked x_1 are open or closed together and hence whether x_2 is open or closed is irrelevant. Indeed the corresponding structure function is $\phi(x_1, x_2) = 1 - (1 - x_1)(1 - x_1 x_2) = x_1$.

Problem 8. If $\phi(x_1, x_2) = x_1[1 - (1 - x_1)(1 - x_2)]$, show that the second component is irrelevant. ∎

5.5 RELIABILITY

We now return to the question of the *reliability* of a system. Since this is the probability that the system is functioning, it is also the probability that the structure function takes the value one. But since this function assumes only the values 0,1, the reliability is further the *expected value of the structure function*. In the case when the components function independently, the reliability, $h(.)$, is a function of the individual component reliabilities, p_i, and, from the structure function, is readily computed. To summarise the notation,

$$h(\mathbf{p}) = h(p_1, p_2, \ldots, p_n) = Pr[\phi(X_1, X_2, \ldots, X_n) = 1]$$
$$= E[\phi(X_1, X_2, \ldots, X_n)].$$

Should all p_i be equal, say to p, then we shall write $h(p)$.

Problem 9. Show for the k out of n system that if the components are independent and each have reliability p, then the system reliability is

$$\sum_{i=k}^{n} \binom{n}{i} p^i (1-p)^{n-i}. \quad \blacksquare$$

Problem 10. System I has reliability $h_I(\mathbf{p}_1)$ and the independent system II has reliability $h_{II}(\mathbf{p}_2)$. Show that

(i)   the resultant reliability when I and II are joined in series is $h_I(\mathbf{p}_1)h_{II}(\mathbf{p}_2)$;

(ii) the resultant reliability when I and II are joined in parallel is $1 - [1 - h_I(\mathbf{p}_1)][1 - h_{II}(\mathbf{p}_2)]$; and

(iii) if each component in system II is replaced by a copy of system I, then the composite reliability is $h_{II}[h_I(\mathbf{p}_1)]$, that is to say the value of the function $h_{II}(.)$ when all its component variables equal $h_I(\mathbf{p}_1)$. $\blacksquare$

We state an interesting result for reliability functions which is due to Moore and Shannon.[1] If a system consists of independent components each with reliability p, and there exists p_0 such that $h(p_0) = p_0$, for $0 < p_0 < 1$, then

$$h(p) < p, \quad 0 \leqslant p < p_0$$
$$h(p) > p, \quad p_0 < p \leqslant 1.$$

It is assumed that $h(p)$ is not in fact identically p and thus the system must contain at least two relevant parts. There are strong intuitive grounds to support this result. For the function $h(.)$ satisfies $h(0) = 0$, $h(1) = 1$. and $0 < h(p) < 1$ and $h(p)$ ought to be increasing in p.

Example 7

For the 2 out of 3 system, when the components are independent and each has reliability p, then $h(p) = 3p^2 - 2p^3$ (see Fig. 5.7).

If $h(p_0) = p_0$, then $3p_0^2 - 2p_0^3 = p_0$,

$$2p_0^2 - 3p_0 + 1 = 0, \text{ if } p_0 \neq 0$$
$$(2p_0 - 1)(p_0 - 1) = 0,$$

and $p_0 = \frac{1}{2}$ is the root which satisfies $0 < p_0 < 1$. It is easily checked that $h(p) < \frac{1}{2}$ for $0 \leqslant p < \frac{1}{2}$ and $h(p) > \frac{1}{2}$ for $\frac{1}{2} < p \leqslant 1$.

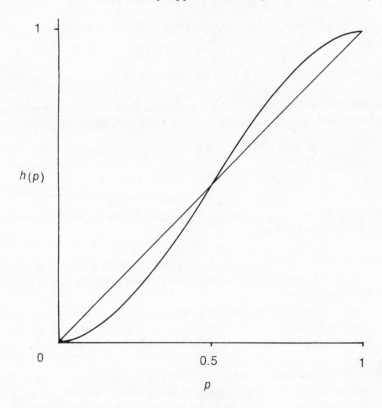

Figure 5.7

5.6 THE LIFE OF A SYSTEM

If we are interested in the time to first failure of a system, then we must re-instate the time variable. If the lives of the individual components are independent continuous random variables with cumulative distribution functions $F_i(t)$ $(i = 1,2, \ldots , n)$, then the probability that the ith component is functioning at time t is $1-F_i(t)$. For a system consisting of n such components in series,

$$h[\mathbf{p}(t)] = \prod_1^n [1-F_i(t)], \tag{5.8}$$

and for n such components in parallel,

$$h[\mathbf{p}(t)] = 1 - \prod_1^n F_i(t). \tag{5.9}$$

Furthermore, for n independent continuous components, the mean time to failure of the system is

$$\int_0^\infty h\,[\mathbf{p}(t)]\,\mathrm{d}t. \tag{5.10}$$

Example 8
A system consists of two independent units in parallel. The life of the first of these is exponentially distributed with parameter λ_1, the second is exponentially distributed with parameter λ_2. Then $F_i(t) = 1-\exp(-\lambda_i t)$, $i = 1,2$. Hence, from equation (5.9),

$$\begin{aligned} h[p_1(t), p_2(t)] &= 1-(1-e^{-\lambda_1 t})(1-e^{-\lambda_2 t}) \\ &= e^{-\lambda_1 t} + e^{-\lambda_2 t} - e^{-(\lambda_1+\lambda_2)t}. \end{aligned} \tag{5.11}$$

The mean time to failure of the system is

$$\int_0^\infty [e^{-\lambda_1 t} + e^{-\lambda_2 t} - e^{-(\lambda_1+\lambda_2)t}]\,\mathrm{d}t = \frac{1}{\lambda_1} + \frac{1}{\lambda_2} - \frac{1}{\lambda_1+\lambda_2}.$$

Problem 11. Find the mean time to failure of a system which consists of two independent components in parallel, the component lives being uniformly distributed over $(0,t_0)$. ■

Problem 12. In a k out of n system, the components are independent and each has reliability $\exp(-\lambda t)$ at time t. Show that the mean life of the system is

$$\sum_{i=k}^{n} 1/i\lambda. \quad ■$$

Problem 13. For Example 8, show that the failure rate (see section 4.12) for the system is

$$\frac{\lambda_1 e^{-\lambda_1 t} + \lambda_2 e^{-\lambda_2 t} - (\lambda_1+\lambda_2)e^{-(\lambda_1+\lambda_2)t}}{e^{-\lambda_1 t} + e^{-\lambda_2 t} - e^{-(\lambda_1+\lambda_2)t}}.$$

Show further that the derivative of the failure rate, with respect to time, is a positive multiple of

$$[\lambda_2^2 e^{-\lambda_1 t} + \lambda_1^2 e^{-\lambda_2 t} - (\lambda_1-\lambda_2)^2].$$

By considering the sign of the derivative at $t = 0$ and when $t \to \infty$, show that, if $\lambda_1 \neq \lambda_2$, the failure rate is first increasing and then decreasing. ■

5.7 BOUNDS FOR RELIABILITY

It will be apparent that if the system is complicated then computing the exact reliability may prove tedious. An outline will now be given of a method of splitting up the system into sections which is particularly useful for computing upper and lower bounds for the reliability.

Suppose then that a system of n components has (binary) structure function $\phi(x)$ where $x = (x_1, x_2, \ldots, x_n)$. A **path vector** is any x such that $\phi(x) = 1$. A path vector x is said to be *minimal* if for every $y < x$, then $\phi(y) = 0$. By $y < x$ we mean $y_i \leqslant x_i$ $(i = 1, 2, \ldots, n)$ with $y_i < x_i$ for at least one i.

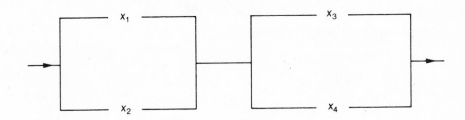

Figure 5.8

Example 9

The structure function for the system shown in Fig. 5.8 is readily computed. It is used to illustrate the ideas of this section. Here $n = 4$ and $x = (x_1, x_2, x_3, x_4)$. Clearly $(1,0,1,0)$ is a path vector but $(1,1,0,0)$ is not a path vector. $(1, x_2, 1, x_4)$ is a path vector whatever the values of x_2, x_4. $(1,0,1,1)$ is a path vector but it is not minimal since $(1,0,0,1) < (1,0,1,1)$ and is itself a path vector. However, $(1,0,0,1)$ is a minimal path vector.

The sense of a minimal path vector is that there is a subset of $x_1, x_2, \ldots, x_n$, say S_j, such that if every $x_i \in S_j$ is 1 and all the rest have the value 0, then the system functions. But whenever *any* $x_i \in S_j$ has value zero, in addition to those not in S_j, then the system does not function. To each S_j we assign a minimal path function

$$\alpha_j(x) = \prod_{x_i \in S_j} x_i, \qquad (5.12)$$

which takes the value 1 if and only if every $x_i = 1$, $x_i \in S_j$.

Example 10

For the system in Example 9, the minimal path vectors are:

$$S_1 \;=\; (1,0,1,0) \;,\; \alpha_1(\mathbf{x}) \;=\; x_1 x_3.$$
$$S_2 \;=\; (1,0,0,1) \;,\; \alpha_2(\mathbf{x}) \;=\; x_1 x_4.$$
$$S_3 \;=\; (0,1,1,0) \;,\; \alpha_3(\mathbf{x}) \;=\; x_2 x_3.$$
$$S_4 \;=\; (0,1,0,1) \;,\; \alpha_4(\mathbf{x}) \;=\; x_2 x_4.$$

The whole system will function if and only if at least one $\alpha_j(\mathbf{x})$ takes the value 1. Hence the S_j are to be thought as 'in parallel' and $\phi(.)$ is expressible as

$$\phi(\mathbf{x}) = 1 - \prod_j [1 - \alpha_j(\mathbf{x})]. \tag{5.13}$$

Example 11
For the system in Example 9, using the results in Example 10,

$$\phi(x_1, x_2, x_3, x_4) = 1 - (1 - x_1 x_3)(1 - x_1 x_4)(1 - x_2 x_3)(1 - x_2 x_4).$$

We now have one obvious lower bound for the reliability of a system. For

$$Pr[\phi(\mathbf{X}) = 1] \;=\; Pr[\text{at least one } \alpha_j(\mathbf{X}) = 1]$$
$$\geqslant Pr[\alpha_j(\mathbf{X}) = 1], \text{ for each } j.$$

Hence

$$Pr[\phi(\mathbf{X}) = 1] \;\geqslant\; \max_j \{ Pr[\alpha_j(\mathbf{X}) = 1] \}. \tag{5.14}$$

Example 12
In Example 9, if the components are independent, each with reliability p, then $Pr[\alpha_j(\mathbf{X}) = 1] = p^2, j = 1,2,3,4$. Hence the reliability of the whole system is at least p^2.

Instead of locating sets of components whose functioning guarantees functioning of the whole system, we can look for sets of components whose failures ensures failure of the whole system. Thus $\mathbf{x}$ is said to be a **cut vector** for the structure function $\phi(.)$ if $\phi(\mathbf{x}) = 0$. A cut vector $\mathbf{x}$ is said to be *minimal* if for every $\mathbf{y} > \mathbf{x}$, then $\phi(\mathbf{y}) = 1$.

Example 13
For Example 9, $(0,0,0,1)$ is a cut vector, but it is not minimal since $(0,0,1,1) > (0,0,0,1)$ and is also a cut vector. $(x_1, x_2, 0, 0)$ is a cut vector whatever x_1, x_2 but is only minimal if $x_1 = x_2 = 1$.

The sense of a minimal cut vector is that there is a subset of $x_1, x_2, \ldots, x_n$, say T_j, such that if every $x_i \in T_j$ is 0 and all the rest have the value 1, then the system fails. But whenever *any* $x_1 \in T_j$ has value 1, in addition to those not in T_j, then the system functions. To each T_j we assign a minimal cut function

$$\beta_j(\mathbf{x}) = 1 - \prod_{x_i \in T_j} (1-x_i), \qquad (5.15)$$

which takes the value 0, if and only if every $x_i = 0, x_i \in T_j$.

Example 14
For Example 9,

$$T_1 = (0,0,1,1) \ , \ \beta_1(\mathbf{x}) = 1-(1-x_3)(1-x_4)$$
$$T_2 = (1,1,0,0) \ , \ \beta_2(\mathbf{x}) = 1-(1-x_1)(1-x_2).$$

The whole system functions if and only if all the $\beta_j(\mathbf{x})$ take the value 1. Hence the T_j are to be thought of as 'in series' and $\phi(.)$ is expressible as

$$\phi(\mathbf{x}) = \prod_j \beta_j(\mathbf{x}). \qquad (5.16)$$

Example 15
For the system in Example 9, using the results in Example 14,

$$\phi(\mathbf{x}) = [1-(1-x_3)(1-x_4)] \ [1-(1-x_1)(1-x_2)].$$

As a problem, the reader should check that this simplifies to the same binary function given in Example 11. Remember that $x_i^2 = x_i$ for binary variables.

We now have an obvious upper bound for the reliability of a system. For,

$$Pr[\phi(\mathbf{X}) = 1] \ = \ Pr[\text{all } \beta_j(\mathbf{X}) = 1]$$
$$\leqslant Pr[\beta_j(\mathbf{X}) = 1], \text{ for each } j.$$

Hence

$$Pr[\phi(\mathbf{X}) = 1] \ \leqslant \ \min_j \{Pr[\beta_j(\mathbf{X}) = 1]\}. \qquad (5.17)$$

Example 16
In Example 9, if the components are independent, each with reliability p, then $Pr[\beta_j(\mathbf{X}) = 1] \ = \ 1-(1-p)^2, j = 1,2$. Hence the reliability of the system is at most $1-(1-p)^2$.

Problem 14

For the system with four components arranged as shown in Fig. 5.9, find the structure function via the minimal path vectors and the minimal cut vectors. If the components function independently and each have reliability p, use equations (5.14) and (5.17) to find lower and upper bounds for the reliability of the system.

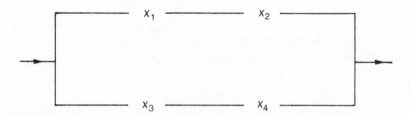

Figure 5.9

*5.8 EFFECT OF CORRELATION

Even if we started out with independent components, the sub-structure functions for minimal path and cut vectors need no longer be independent. Supposing we ignore the strain caused on components arising from the failure of some, it remains the case that a component may belong to more than one minimal path vector. Hence if we know that a component is functioning then this increases the probability that the sub-systems to which it belongs are functioning. Thus if Y_1, Y_2 are binary random variables such that $Y_i = 1$ if and only if sub-system i is functioning $(i = 1,2)$, then

$$Pr(Y_1 = 1 \text{ and } Y_2 = 1)$$
$$= Pr(Y_2 = 1 | Y_1 = 1)Pr(Y_1 = 1)$$
$$\geqslant Pr(Y_2 = 1)Pr(Y_1 = 1). \tag{5.18}$$

Similarly if a sub-system fails, this will increase the probability that another sub-system with which it has components in common will also fail. Hence

$$Pr(Y_1 = 0 \text{ and } Y_2 = 0)$$
$$= Pr(Y_2 = 0 | Y_1 = 0)Pr(Y_1 = 0)$$
$$\geqslant Pr(Y_2 = 0)Pr(Y_1 = 0). \tag{5.19}$$

We first apply these results to equation (5.13).

$$Pr[\phi(\mathbf{X}) = 1] = Pr\left\{1 - \prod_j [1-\alpha_j(\mathbf{X})] = 1\right\},$$

$$= Pr[\text{at least one } \alpha_j(\mathbf{X}) = 1],$$

$$= 1 - Pr[\text{all } \alpha_j(\mathbf{X}) = 0], \text{ and from equation (5.19)}$$

$$\leqslant 1 - \prod_j Pr[\alpha_j(\mathbf{X}) = 0],$$

$$= 1 - \prod_j \{1 - Pr[\alpha_j(\mathbf{X}) = 1]\}. \tag{5.20}$$

Example 17
For Example 9, if the components are independent and each have reliability p, then, from Example 12, $Pr[\alpha_j(X) = 1] = p^2$. Hence the reliability of the whole system is at most $1 - (1-p^2)^4$, from equation (5.20).

Similarly we consider equation (5.16), in virtue of equation (5.18),

$$Pr[\phi(X) = 1] = Pr\{\prod_j [\beta_j(X) = 1]\}$$

$$\geqslant \prod_j \{Pr[\beta_j(X) = 1]\}. \tag{5.21}$$

Example 18
For Example 9, if the components are independent and have reliability p, then from Example 16, $Pr[\beta_j(X) = 1] = 1-(1-p)^2$. Hence the reliability of the system is at least $[1-(1-p)^2]^2$. In fact, the reliability of the system *is* $[1-(1-p)^2]^2$ and thus the lower bound is attained. This is because the two minimal cut vectors have no component values in common.

Problem 15. Compare the upper and lower bounds in Examples 17 and 18 with those found in Examples 12 and 16. ∎

Problem 16. Use equations (5.20) and (5.21) to determine upper and lower bounds for the reliability of the system in Problem 14. ∎

REFERENCE

[1]   'Reliable circuits using less reliable relays', B. F. Moore and C. E. Shannon. *Journal of the Franklin Institute* (1956).

BRIEF SOLUTIONS AND COMMENTS ON THE PROBLEMS

Problem 1. $Pr[(A \text{ and } B) \text{ or } (C \text{ and } D)] = p^2 + p^2 - (p^2)(p^2) = p^2(2-p^2)$. Alternatively as $1-(1-p^2)^2$. A,B in series, C,D in series, while the pair A,B are in parallel with the pair C,D.

Problem 2. $[1-(1-p)^2]^2 > p$ if $(1-p)[1-2(1-p) + (1-p)^3]$
$= (1-p)p(-1+3p-p^2) > 0$. That is $p^2 - 3p + 1 < 0 \Rightarrow p > (3-\sqrt{5})/2 \simeq 0.38$. The system in Example 1 will be more reliable than that of Problem 2 if

$$[1-(1-p)^2]^2 > 1-(1-p^2)^2$$

or $(1-p)^2 (1+p)^2 - 2(1-p)^2 + (1-p)^4 > 0$

$\Leftrightarrow$ $(1+p)^2 - 2 + (1-p)^2 > 0$

$\Leftrightarrow$ $2p^2 > 0$, true for all $p > 0$.

Problem 3. Denote (A,C), (A,D), (B,C), (B,D), by 1,2,3,4 respectively. Let p_i be the probability that both members of pair i are functioning, p_{ij} that both members of pairs i and j are functioning – and so on.

$$p_1 = p_2 = p_3 = p_4 = p^2, p_{12} = p^3, p_{13} = p^3, p_{14} = p^4.$$

$$p_{23} = p^4, p_{24} = p^3, p_{34} = p^3, p_{123} = p_{124} = p_{234} = p_{134} = p^4,$$

$$p_{1234} = p^4.$$

Using formula $\sum_i p_i - \sum \sum_{i \neq j} p_{ij} + \sum \sum \sum_{i \neq j \neq k} p_{ijk} - \sum \sum \sum \sum_{i \neq j \neq k \neq \ell} p_{ijk\ell}$, the probability of at least one pair functioning is

$$4p^2 - (4p^3 + 2p^4) + 4p^4 - p^4.$$

Problem 4
(i) If any $\quad x_i = 0$, then min $(x_1, x_2, \ldots, x_n) = 0$,
    if every $\quad x_i = 1$, then min $(x_1, x_2, \ldots, x_n) = 1$.
(ii) If at least one $x_i = 1$, then max $(x_1, x_2, \ldots, x_n) = 1$,
    if every $\quad\quad x_i = 0$, then max $(x_1, x_2, \ldots, x_n) = 0$.

Problem 5

$$
\begin{aligned}
Pr[\max(T_1, T_2) \geqslant t] &= 1 - Pr[\max(T_1, T_2) < t] \\
&= 1 - Pr(T_1 < t \text{ and } T_2 < t) \\
&= 1 - Pr(T_1 < t)Pr(T_2 < t),
\end{aligned}
$$

since T_1, T_2 are independent. The result follows. $Pr[\min(T_1, T_2) \geqslant t] = Pr[T_1 \geqslant t \text{ and } T_2 \geqslant t] = Pr(T_1 \geqslant t)Pr(T_2 \geqslant t)$.

Problem 6. A,B are in series, the structure function is $x_1 x_2$. C,D are in series, the structure function is $x_3 x_4$. But these two pairs are in parallel, hence

$$\phi(x_1, x_2, x_3, x_4) = 1 - (1 - x_1 x_2)(1 - x_3 x_4).$$

Problem 7. The expression lists 'at least 2 out of 3' by displaying the possibilities for just two and for all three.

Problem 8

$$\phi(1,1) = \phi(1,0) = 1, \phi(0,0) = \phi(0,1) = 0.$$

Hence x_2 is irrelevant. Alternatively,

$$
\begin{aligned}
x_1[1 - (1 - x_1)(1 - x_2)] &= x_1(x_1 + x_2 - x_1 x_2) = x_1^2 + x_1 x_2 - x_1^2 x_2 \\
&= x_1 + x_1 x_2 - x_1 x_2 = x_1 \text{ since } x_1^2 = x_1.
\end{aligned}
$$

Problem 9. There are $\binom{n}{i}$ different selections of i components to function and $(n-i)$ which are not to function. The probability is $p^i(1-p)^{n-i}$ for each such selection.

Problem 10

(i)   Since the sub-systems are in series, the composite system will function if and only if system I and system II both function.

(ii) Since the sub-systems are in parallel, the composite system will function if and only if at least one of systems I and II functions.

(iii) The ith element of $\mathbf{p}_2$ is the probability that the ith component is functioning. If the ith component is replaced by a copy of system I, then the probability that it functions is $h_I(\mathbf{p}_1)$. For example, if $h_{II}(p) = p^2$ and $h_I(p) = 1-(1-p)^2$, then $h_{II}[h_I(p)] = [1-(1-p)^2]^2$. The reader should draw a sketch of systems I and II for this example.

Problem 11. If T_1, T_2 are the lives of the individual components, $Pr(T_1 < t$ and $T_2 < t) = Pr(T_1 < t)Pr(T_2 < t) = (t/t_0)\,(t/t_0) = (t/t_0)^2$. Hence p.d.f. of failure time of $\max(T_1, T_2)$ is $\dfrac{\mathrm{d}}{\mathrm{d}t}\,[(t/t_0)^2] = 2t/t_0^2$. Expected time is

$$\int_0^{t_0} (2t^2/t_0^2)\mathrm{d}t = 2t_0/3.$$

Alternatively, $h[p_1(t), p_2(t)] = 1-(t/t_0)^2$ and $\displaystyle\int_0^{t_0} [1-(t/t_0)^2]\,\mathrm{d}t = 2t_0/3$.

Problem 12. If at any time, i are functioning, then the time to the next failure is the minimum of the i exponentially distributed lives and has expectation $1/i\lambda$. The system fails as soon as there have been $n-k+1$ failures. Hence time to the first failure has mean $1/n\lambda$, the *additional* time to the second has mean $1/[(n-1)\lambda]$ and so on. Alternatively consider $\displaystyle\int_0^\infty h(t)\mathrm{d}t$. where $h(t) = \displaystyle\sum_{r=k}^{n} \binom{n}{r}$ $\exp(-r\lambda t)\,[1-\exp(-\lambda t)]^{n-r}$.

Problem 13. Writing the failure rate as

$$\frac{\lambda_1\,\exp(\lambda_2 t) + \lambda_2\,\exp(\lambda_1 t) - (\lambda_1+\lambda_2)}{\exp(\lambda_2 t) + \exp(\lambda_1 t) - 1} = \frac{u}{v},$$

differentiating the quotient and collecting terms, we have,

$$\{\lambda_2^2\,\exp(\lambda_2 t) + \lambda_1^2\,\exp(\lambda_1 t) - (\lambda_2-\lambda_1)^2\,\exp[(\lambda_1+\lambda_2)t]\}/v^2.$$

The value at $t = 0$ is $\lambda_2^2 + \lambda_1^2 - (\lambda_2-\lambda_1)^2 = 2\lambda_1\lambda_2 > 0$, but the value is negative for sufficiently large t provided $\lambda_1 \neq \lambda_2$.

Problem 14

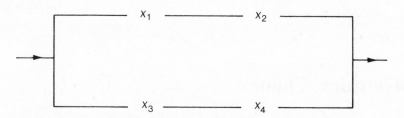

Figure 5.10

The minimal path vectors are $S_1 = (1,1,0,0)$, $S_2 = (0,0,1,1)$. The corresponding minimal path functions are $\alpha_1(x) = x_1 x_2$, $\alpha_2(x) = x_3 x_4$. Hence, from (5.13), $\phi(x) = 1-(1-x_1 x_2)(1-x_3 x_4)$. The minimal cut-vectors are

$$T_1 = (0,1,0,1), \qquad \beta_1(x) = 1-(1-x_2)(1-x_4).$$
$$T_2 = (0,1,1,0), \qquad \beta_2(x) = 1-(1-x_2)(1-x_3).$$
$$T_3 = (1,0,0,1), \qquad \beta_3(x) = 1-(1-x_1)(1-x_4).$$
$$T_4 = (1,0,1,0), \qquad \beta_4(x) = 1-(1-x_1)(1-x_3).$$

$\phi(x)$ is obtained from (5.16).

$Pr[\alpha_i(X) = 1] = p^2, i = 1,2$. From (5.14), $Pr[\phi(X) = 1] \geqslant p^2$.

$Pr[\beta_i(X) = 1] = 1-(1-p)^2, i = 1,2,3,4$. From (5.17),

$Pr[\phi(X) = 1] \leqslant 1-(1-p)^2$.

Problem 15. The system is that of Example 9. From Examples 12 and 16, lower and upper bounds are p^2, $1-(1-p)^2$. From Examples 18 and 17, lower and upper bounds are $[1-(1-p)^2]^2$, $1-(1-p^2)^4$.

Now $[1-(1-p)^2]^2 - p^2 = (2p-p^2)^2 - p^2 = p^2[(2-p)^2-1] = p^2(1-p)(3-p) > 0$, whereas

$1-(1-p^2)^4 - [1-(1-p)^2] = (1-p)^2 [1-(1-p)^2 (1+p)^4]$
$= (1-p)^2 [1-(1-p^2)^2 (1+p)^2] = (1-p)^2 p[1+(1-p^2)(1+p)] (p^2+p-1)$ and
p^2+p-1 changes sign in the interval $0 < p < 1$.

Problem 16

$$[1-(1-p)^2]^4 \leqslant Pr[\phi(X) = 1] \leqslant 1-(1-p^2)^2.$$

In fact the upper bound is attained, this is because the minimal path vectors have no component values in common.

CHAPTER 6

Inventory Theory

6.1 INTRODUCTION

An inventory is a stock of items or material stored for use in the future. The maintenance of an inventory implies a replacement policy and incurs costs of several different kinds. There is, to begin with, an **ordering cost**, which for y items is typically of the form $k + c(y)$. Here, k is a fixed charge for placing any positive order, and the function $c(y)$ may reflect any advantages obtainable from bulk buying.

Then there are storage or **holding costs**, which might include the loss of interest on frozen capital, the renting of space, and insurance. Such costs are often assumed to be a function of either the average stock held during one accounting cycle or of the remaining stock at the end of such a period.

Shortage or **penalty costs** may be incurred when demand exceeds supply. If customers will not wait **(no backlogging)**, then either potential sales are lost or their demands will have to be met by special deliveries at enhanced rates. Even if customers are prepared to wait **(backlogging allowed)**, there may be additional administrative costs, not to mention a possible loss of goodwill.

The main source of revenue arises from sales but sometimes the stock remaining at the end of a period has a salvage value. The excess of revenue over costs represents the profit of the operation. The concern of management is to maximise profit but this cannot be engineered solely by increasing prices since demand may thereby be depressed. However, management does have control over the re-stocking policy which determines the amount and frequency of re-orders. This aspect emphasises the role of costs and suggests that the economic performance of a policy be assessed by the loss incurred. The *loss* is defined as being the cost less the revenue. Evidently a negative loss corresponds to a positive profit and the equivalent optimisation problem is to minimise the expected loss. To begin with we look briefly at a deterministic problem, in which there is a constant rate of demand for items.

Example 1

A store is initially empty. There is a fixed cost, k, for placing an order of any size and a charge of c per unit amount of a material which is to be stored to meet future demand. Thus if the amount ordered is $y(>0)$, then the ordering costs is of the form $k + cy$. We shall assume that orders are delivered immediately, that there is a constant demand for an amount m per unit time, and that the sale of a unit yields a revenue of r. It is not permitted to run out of stock so no shortage charge can arise. There is, however, a holding charge of h per unit of time spent by a unit quantity in the store. It is required to minimise the loss per unit of time.

If the store starts with y, then at a time t later it falls to $y-mt$ $(\geqslant 0)$. The average quantity in the store is $\frac{1}{2}(y + y-mt)$ and hence the holding charge is $\frac{h}{2}(2y-mt)t$. Subtracting the revenue, rmt, the total loss over $(0,t)$ is

$$k + cy - rmt + h(2y-mt)t/2. \tag{6.1}$$

Hence the loss per unit time is

$$\frac{k}{t} + \frac{cy}{t} - rm + \frac{h(2y-mt)}{2}. \tag{6.2}$$

The expression in (6.2) increases with y and hence is a minimum when $y = mt$ (recall that the store is not allowed to run out of material). This confirms that it is not worth ordering again until the store is empty. Since we are primarily interested in the amount to be ordered we set $t = y/m$ in (6.2) and obtain

$$\frac{mk}{y} + m(c-r) + \frac{hy}{2}. \tag{6.3}$$

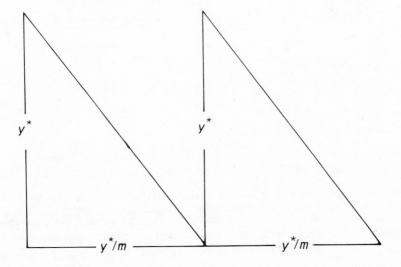

Figure 6.1

After differentiating with respect to y, it is easily shown that (6.3) has a minimum at $y^* = \sqrt{2km/h}$, the so-called **economic lot size**. The reader should envisage a sequence of re-orders at regular intervals $y^*/m = \sqrt{2k/mh}$ (see Fig. 6.1).

The same optimum amount can be ordered if there is a delay in delivery but this is of fixed duration ($< y^*/m$). Early orders must be placed to ensure deliveries at times ny^*/m ($n = 1,2,\ldots$).

Problem 1. If in Example 1, $k = 9$, $c = 1$, $h = 1$, $r = 5$, $m = 2$ calculate the economic lot size and hence the corresponding total loss for 12 units of time. Use (6.1) to find the lot size for which the total loss over $t = y/m$ is a minimum and the corresponding loss for 12 units of time. Comment on the two results. ∎

Example 2
We consider modifying Example 1 to allow a shortage of stock, the excess demand being backlogged and satisfied at the next re-order time. A graph of the

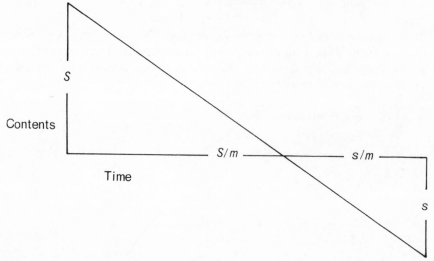

Figure 6.2

contents against time is shown in Fig. 6.2. We suppose that a backlog of s is allowed before ordering $S+s$, which raises the contents of the store to S. The store runs out after a time S/m, so that the holding charge is $\frac{1}{2} S \left(\frac{S}{m}\right) h$. The backlog introduces a new feature, namely a shortage penalty. Suppose this is exacted at p per unit time that a unit demand is outstanding. Then, since the store is empty for a time s/m and the maximum shortage is s, the penalty is $\frac{1}{2} s \left(\frac{s}{m}\right) p$. The total loss for the duration $(S+s)/m$ between re-orders is

$$k + c(S+s) + \frac{hS^2}{2m} + \frac{ps^2}{2m} - r(S+s). \tag{6.4}$$

Hence the loss per unit time is

$$\frac{mk}{S+s} + \frac{hS^2}{2(S+s)} + \frac{ps^2}{2(S+s)} + (c-r)m. \tag{6.5}$$

The expression in (6.5) has to be minimised. By taking the partial derivatives with respect to S,s and equating to zero it is easily shown that the optimal values, S^*,s^* satisfy $s^* = (h/p)S^*$. After substituting for s^* in one of the equations we eventually arrive at $S^* = \sqrt{2mkp/[h(h+p)]}$. We remark that as p increases, then $s^* \to 0$ and $S^* \to \sqrt{2mk/h}$, agreeing with the result obtained in Example 1 when running out of stock was not allowed.

Problem 2. For Example 2, we may make a different assumption about the shortage penalty, namely that a charge of p is levied for each unit backlogged. Show that, if the other charges remain the same, the loss per unit time is

$$\frac{2km + hS^2 - 2pmS}{2(S+s)} + m(p+c-r).$$

Show further that if $p > \sqrt{2kh/m}$, the loss per unit time is minimised when $s = 0$, $S = \sqrt{2km/h}$. (Hint: first find an interval for S for which $2km+hS^2 - 2pmS < 0$ and note that, for such S, the loss per unit time decreases as s decreases.) ∎

6.2 RANDOM DEMAND

The problems that really interest us, however, are those in which the demand is random and the amount of stock in deficit or in surplus at some later time cannot be guaranteed in advance. We first discuss the so-called newspaper problem.

Example 3

A newspaper-seller buys x papers at the beginning of each day at a cost of c pence each and sells them at $r > c$ pence each. Any papers unsold at the end of the day are an outright loss, and nothing can be done to satisfy demand in excess of x. How should x be chosen so as maximise his expected profit? If the demand is for Z papers, then his revenue is rZ if $Z \leqslant x$, but is rx if $Z > x$, and his costs are cx in either case. Hence his *expected* profit per day is

$$r \sum_{z=0}^{x} zp(z) + rx \sum_{z=x+1}^{\infty} p(z) - cx, \tag{6.6}$$

where $p(z)$ is the probability that $Z = z$. If the order is increased to $x+1$, the expected profit is

$$r \sum_{z=0}^{x+1} zp(z) + r(x+1) \sum_{z=x+2}^{\infty} p(z) - c(x+1).$$

Subtracting the expected profit for x from the expected profit for $x+1$, we have

$$r(x+1)p(x+1) + rx \sum_{z=x+2}^{\infty} p(z) + r \sum_{z=x+2}^{\infty} p(z) - rx \sum_{z=x+1}^{\infty} p(z) - c$$

$$= rp(x+1) + r \sum_{z=x+2}^{\infty} p(z) - c = r \sum_{z=x+1}^{\infty} p(z) - c$$

$$= r\,Pr(Z \geqslant x+1) - c.$$

Hence $x+1$ should be ordered in preference to x so long as $Pr(Z \geqslant x+1) > c/r$. But $Pr[Z \geqslant x+1]$ is decreasing as x increases. Hence the expected profit is maximised by ordering x^* such that x^* is the greatest integer satisfying $Pr[Z \geqslant x^*] > c/r$. For example, if $p(z) = (1-\alpha)\alpha^z$, $z = 0,1,2, \ldots$, then $Pr(Z \geqslant x^*) = \alpha^{x^*}$ and x^* is the greatest integer such that $\alpha^{x^*} > c/r$.

The task of finding the maximum (or minimum) of a function is much simplified if calculus methods are employed. To make these available we must treat the stock as a continuous quantity. This practice is indeed recognised when small items are sold by weight. The demand must also be approximated by a continuous random variable. The reader will already have met, for example, the use of a normal approximation to the Poisson distribution (see Johnson and Kotz[1]).

We now return to Example 3 and assume that Z has a continuous distribution with p.d.f. $f(z)$. The equation equivalent to (6.6) is

$$r \int_0^x zf(z)dz + rx \int_x^{\infty} f(z)dz - cx. \qquad (6.7)$$

The derivative of (6.7) with respect to x is

$$rxf(x) + r \int_x^{\infty} f(z)dz - rxf(x) - c$$

$$= r \int_x^{\infty} f(z)dz - c. \qquad (6.8)$$

The derivative of (6.8) is $-rf(x) < 0$. Hence (6.7) has a maximum at x^*, where, from (6.8),

$$\int_{x^*}^{\infty} f(z)dz = c/r \text{ or}$$

$$Pr(Z \geqslant x^*) = c/r.$$

We shall adopt the technique of approximating the demand by a continuous distribution for most of the rest of this chapter.

Problem 3. Suppose in Example 3, that all revenue actually received is disregarded but, demand that cannot be satisfied is regarded as a cost. Show that the value of x which minimises the expected cost satisfies $Pr(Z \geqslant x^*) = c/r$. ■

Problem 4. Show for the newspaper problem, that if a salvage value of s per unsold copy is allowed and demand is approximated by a continuous random variable, then x^* satisfies $Pr(Z \geqslant x^*) = (c-s)/(r-s)$; one can risk stocking more since $c < r$. ■

One feature in our example is rather special. Since a newspaper has value only on the day of issue, stock is not carried over and our decision as to the optimum inventory is the same for each day, in so far as the conditions remain the same.

6.3 OPTIMUM INVENTORY FOR ONE PERIOD

The store already contains x items. At the beginning of a single period, of fixed duration, it is possible to order additional items which, we suppose, are delivered immediately. At the end of the period there is a holding charge based on the number of items then remaining. If the stock runs out there is a penalty based on the number of unsatisfied demands, which are regarded as lost sales. Each demand is for a single item but we approximate the distribution of demands by a continuous probability density function. The question is: to order or not to order? Moreover, if an order be placed, to what level should the inventory be raised to minimise the expected loss for the period?

Linear costs

Suppose the cost is c for each item ordered, but there is no additional fixed charge for placing an order. The holding charge is h per item in surplus and the shortage penalty p per unsatisfied demand, both levied at the end of the period. Thus if the inventory is brought up to y by ordering $y - x$ items, then the cost is $c(y - x)$. If the demand, Z, is less than y then the holding cost is $h(y - Z)$, but if greater than y, there is a shortage cost of $p(Z - y)$. Thus the expected cost of ordering up to y, starting with x on hand is

$$c(y-x) + p \int_y^\infty (z-y)f(z)\mathrm{d}z + h \int_0^y (y - z)f(z)\mathrm{d}z. \qquad (6.9)$$

If the revenue is r per item, then when the demand is less than y the revenue is rZ but if greater than y, the revenue is limited to ry. Hence the expected revenue is

$$r \int_0^y zf(z)\mathrm{d}z + ry \int_y^\infty f(z)\mathrm{d}z.$$

The expected loss = expected cost − expected revenue, so, writing $\ell(y,x)$ as the expected loss,

$$\ell(y,x) = c(y-x) + (p+r) \int_y^\infty (z-y)f(z)\mathrm{d}z + h \int_0^y (y-z)f(z)\mathrm{d}z - rE(Z). \quad (6.10)$$

We see that, apart from the constant term, $-rE(Z)$, the right-hand side of equation (6.10) has the same form as (6.9), except that the penalty term incorporates an allowance for lost sales. The constant term, $rE(Z)$, does not effect the value of y at which $\ell(y,x)$ attains a minimum. We are to minimise $\ell(y,x)$ *over those $y \geqslant x$.*

Differentiating integrals with respect to a variable which appears both in the limits and in the integrand needs a little care. A selection of useful results in this connection is given at the end of this section. In the linear case, traps may be avoided by splitting into separate integrals. Thus

$$\int_y^\infty (z-y)f(z)\mathrm{d}z = \int_y^\infty zf(z)\mathrm{d}z - y \int_y^\infty f(z)\mathrm{d}z,$$

and the derivative of this expression with respect to y is

$$-yf(y) - \int_y^\infty f(z)\mathrm{d}z + yf(y) = - \int_y^\infty f(z)\mathrm{d}z.$$

Thus the partial derivative of $\ell(y,x)$ in (6.10) with respect to y is

$$c - (p+r) \int_y^\infty f(z)\mathrm{d}z + h \int_0^y f(z)\mathrm{d}z,$$

$$= c - (p+r) + (h+p+r) \int_0^y f(z)\mathrm{d}z. \quad (6.11)$$

The expression in (6.11) is an increasing function of y. A sensible situation requires that $c < (p+r)$ for otherwise, when the store is initially empty, it will not pay to order any stock! Granted this condition, there is a minimum at y^*, which is attainable if x does not already exceed y^*, and

$$Pr(Z \leqslant y^*) = (p+r-c)/(p+r+h). \quad (6.12)$$

So if $x < y^*$, $\ell(y,x)$ has a minimum for $y \geqslant x$ at $y = y^*$. However, if $x \geqslant y^*$, $\ell(y,x)$ is minimised when $y = x$. The optimum policy is thus:

If $x \geqslant y^*$, order nothing, if $x < y^*$, order $(y^* - x)$.

Hence $\ell(x) = \min_{y \geqslant x} \ell(y,x)$ is a function of x and takes the form

$$\ell(x) = \ell(y^*,x), \, x < y^* \tag{6.13}$$

$$\ell(x) = \ell(x,x), \, x \geqslant y^*. \tag{6.14}$$

Example 4

Suppose the demand, Z, has a uniform distribution over $(0,\theta)$. Then $Pr(Z \leqslant y) = y/\theta$. From equation (6.12), y^* must satisfy $y^*/\theta = (p+r-c)/(p+r+h)$. We can find the expected loss under the optimal policy. From (6.10),

$$\ell(y,x) = c(y-x) + (p+r) \int_y^\theta \frac{(z-y)}{\theta} \, dz + h \int_0^y \frac{(y-z)}{\theta} \, dz - rE(Z)$$

$$= c(y-x) + (p+r)\frac{(\theta-y)^2}{2\theta} + \frac{hy^2}{2\theta} - \frac{r\theta}{2}.$$

Whence

$$\ell(x) = \ell(y^*,x) = c(y^*-x) + \frac{(p+r)(\theta-y^*)^2}{2\theta} + \frac{h(y^*)^2}{2\theta} - \frac{r\theta}{2}, \quad 0 < x < y^*$$

$$\ell(x) = \ell(x,x) = \frac{(p+r)(\theta-x)^2}{2\theta} + \frac{hx^2}{2\theta} - \frac{r\theta}{2}, \quad y^* \leqslant x \leqslant \theta.$$

We note that $\ell(x)$ decreases linearly until x reaches y^* and eventually increases after x reaches $\dfrac{p+r}{p+r+h}\theta$.

Problem 5. Show from (6.13) and (6.14) that $\ell'(x) = -c, \, x < y^*$, $\ell'(x) = -(p+r) + (h+p+r) \int_0^x f(z)dz, \, x \geqslant y^*$. ■

Problem 6. If the ordering costs and the revenue remain linear, but the holding and shortage costs are increasing convex functions of their arguments, show that there exists y^* such that, if $x < y^*$, the inventory should be raised to y^*, and, if $x \geqslant y^*$, nothing should be ordered. ■

Problem 7. Show, for linear costs and revenue,

$$\ell(y,0) = -(p+r-c)y + (h+p+r) \int_0^y F(z)dz + pE(Z).$$

Hence show that if $c < p+r$, then for $y < y^*$, $\ell(y,0) < pE(Z)$ where $F(y^*) = (p+r-c)/(h+p+r)$ and comment on the result. ■

In this section, the expected loss when an initial stock of x is raised to y, has taken the form

$$\ell(y,x) = -cx + G(y).$$

In the case when the holding and penalty costs are linear as is the revenue, then if a certain condition obtains then $G(y)$ has a unique minimum. Such a result is useful in practice since it leads to a simple ordering policy, namely that the inventory should be raised to a fixed level unless this level is already exceeded by the stock on hand. Moreover, considerable computational advantage arises from the knowledge that we have to search for only one zero of $G'(y)$. This desirable property is retained in the more general case when the holding and penalty functions are convex (see Problem 6). We return briefly to the linear case to re-emphasise its key features.

From (6.11), if $c \geqslant p+r$, then $G(y)$ is not only increasing but since $G''(y) > 0$, it is never worth ordering any additional amount. If, however, $c < p+r$, then $G'(0) < 0$, $\lim\limits_{y \to \infty} G'(y) = c+h > 0$; and, since $G''(y) > 0$, $G'(y)$ has only one change of sign and so $G'(y) = 0$ gives a unique minimum. We digress for a moment to express sympathy with any readers who would go further and require $c < r$. For what is the merit of a transaction in which the cost of an item exceeds any revenue it can produce? Indeed if we look at the actual, rather than expected losses, these, for an initially empty store, are

$$cy^* + p(z-y^*) - ry^*, \quad z > y^*$$

$$cy^* + h(y^*-z) - rz, \quad z < y^*.$$

If $c > r$, the loss is positive.

Returning to our theme, the convexity of $G(y)$ is sufficient to guarantee that there is at most one zero of $G'(y)$, while the other conditions for small and large value of y, show that $G'(y)$ has at least one zero. It is this property of a single change of sign of $G'(y)$ which is the key. The study of more general holding and penalty functions involves going outside the level of this text. Still, to illustrate some of the points that can arise, we give two further examples.

Example 5

Stock can be purchased at c per unit quantity. Revenue is to be disregarded. There is no holding cost but a fixed penalty p is exacted at the end of the period if there is any shortage. That is to say, the expected loss is

$$\ell(y,x) = c(y-x) + p \int_y^\infty f(z)\mathrm{d}z = -cx + G(y).$$

Suppose the p.d.f. of the demand for one period is $f(z) = 1-z$, $0 \leqslant z \leqslant 1$, $f(z) = (z-1)$, $1 < z \leqslant 2$, $f(z) = 0$ otherwise; $c = 1$, $p = 2$. Then it is easily verified that

$$G(y) = y^2 - y + 2, \qquad 0 \leqslant y \leqslant 1,$$

$$G(y) = -y^2 + 3y, \qquad 1 < y \leqslant 2.$$

$G(y)$ has a minimum at $y = \frac{1}{2}$, a maximum at $y = 3/2$ and $G(1) = G(2) = +2$. Hence the optimum policy is

if $x \leqslant \frac{1}{2}$,	order $\frac{1}{2} - x$,
if $\frac{1}{2} < x \leqslant 1$,	order nothing,
if $1 < x \leqslant 2$,	order $2 - x$.

The reader should confirm that $\ell(x) = \min\limits_{y \geqslant x} \ell(y, x)$ takes the form

$\ell(\frac{1}{2}, x) = -x + 7/4,$	$0 \leqslant x \leqslant \frac{1}{2},$
$\ell(x, x) = x^2 - 2x + 2,$	$\frac{1}{2} < x \leqslant 1,$
$\ell(2, x) = -x + 2,$	$1 < x \leqslant 2,$

and that although $\ell(.)$ is continuous, its derivative is not.

Example 6

If the model in Example 5 is extended to include a revenue of $r > c$ per unit quantity sold, then

$$\ell(y, x) = c(y - x) + p \int_y^\infty f(z) dz - r \int_0^y z f(z) dz - ry \int_y^\infty f(z) dz,$$

$$= -cx + G(y).$$

Then $G'(y) = c - pf(y) - r[1 - F(y)]$, where $f(.), F(.)$ are the p.d.f. and c.d.f. of the demand for one period. Suppose in particular that $f(z) = z \exp(-z)$, then $1 - F(z) = (1 + z) \exp(-z)$. It is easily verified that $G'(y) = c - [(p+r)y + r]\exp(-y)$. We have; $G'(0) = c - r < 0$, $\lim\limits_{y \to \infty} G'(y) = c > 0$ and $\exp(y)$, $[(p+r)y + r]/c$ have one, and only one, value in common. Hence $G'(y)$ has only one change of sign. This is a case where $G''(y)$ changes sign also — its zero corresponding to a point of inflection in the graph of $G(y)$. Hence there exists a y^* such that if $x < y^*$ the inventory is raised to y^* and if $x \geqslant y^*$, nothing is ordered.

The last two examples have shown clearly that the optimum ordering policy depends both on the form of the holding and shortage cost functions *and* on the distribution of demand. A full discussion has been given by Arrow, Karlin and Scarf[2]. We conclude this section with a few calculus results which may be useful in this chapter.

(1) $\quad \dfrac{d}{du} \displaystyle\int_0^u \phi(v,u)dv = \int_0^u \dfrac{\partial\phi(v,u)}{\partial u} \, dv + \phi(u,u),$

(2) $\quad \dfrac{d}{du} \displaystyle\int_u^k \phi(v,u)dv = \int_u^k \dfrac{\partial\phi(v,u)}{\partial u} \, dv - \phi(u,u).$ In particular we have,

(3) $\quad \dfrac{d}{du} \displaystyle\int_0^u \psi(u-v)\phi(v)dv = \int_0^u \psi'(u-v)\phi(v)dv + \psi(0)\phi(u),$ $\qquad$ (6.15)

(4) $\quad \dfrac{d}{du} \displaystyle\int_u^\infty \psi(v-u)\phi(v)dv = - \int_u^\infty \psi'(v-u)\phi(v)dv - \psi(0)\phi(u).$ $\qquad$ (6.16)

In many applications of (6.15) and (6.16), $\psi(0) = 0$.

6.4 ONE PERIOD: EFFECT OF A FIXED ORDERING COST

We reconsider the example of linear costing but with one apparently slight but, as it turns out, non-trivial modification. Suppose that when an order is placed then, apart from the cost per item, an additional fixed cost, k, is imposed. This is sometimes called the *set-up* cost, since the order may imply that a manu-facturing process has to be re-started. Then the expected cost for raising the inventory to $y(> x)$ is

$$\ell(y,x) \;=\; k+c(y-x)+h \int_0^y (y-z)f(z)dz+(p+r)$$

$$\int_y^\infty (z-y)f(z)dz-rE(Z) \qquad (6.17)$$

$$=\; k+c(y-x) + L(y)-rE(Z), \text{ say.} \qquad (6.18)$$

On the other hand, if nothing is ordered, there can be no buying cost and

$$\ell(x,x) \;=\; h \int_0^x (x-z)f(z)dz+(p+r) \int_x^\infty (z-x)f(z)dz-rE(Z)$$

$$=\; L(x) - rE(Z).$$

The derivative of $\ell(y,x)$ with respect to y is

$$(c-p-r) + (h+p+r) \int_0^y f(z)dz$$

and there is a minimum at y^*, where

$$Pr(Z \leqslant y^*) = (p+r-c)/(p+r+h), \qquad (6.19)$$

and, provided that $y^* > x$,

$$\ell(y^*,x) = k + c(y^*-x) + L(y^*) - rE(Z).$$

Now we are to minimise $\ell(y,x)$ subject to $y \geqslant x$. So if $x \geqslant y^*$, the position is as before — ordering nothing. If, however, $x < y^*$, the situation is not immediately clear — owing to the presence of the extra term $k(>0)$ which appears in $\ell(y,x)$, $y > x$, but not in $\ell(x,x)$. So for $x < y^*$, there may be values of x such that

$$L(x) < k + c(y^* - x) + L(y^*):$$

that is

$$cx + L(x) < k + cy^* + L(y^*).$$

But if $x < y^*$, $cx + L(x) > cy^* + L(y^*)$ because y^* minimises (6.18). Hence we find x^* such that

$$cx^* + L(x^*) = k + cy^* + L(y^*). \tag{6.20}$$

If $x^* \leqslant x \leqslant y^*$ we *still* order nothing. If $x < x^*$, we raise the inventory to y^* — i.e. order $y^* - x$. The situation is readily grasped by a rough sketch of $cy + L(y)$.

We have our first example of what is often called an (s,S) policy, namely:

At a re-order point, if the inventory is not less than s, order nothing. If the inventory is less than s, then order so that it is raised to S.

Example 7

Suppose the demand is taken to have an exponential distribution with parameter λ. Then, the linear holding and penalty function is, from (6.18),

$$L(y) = h \int_0^y (y-z)\lambda e^{-\lambda z} \, dz + (p+r) \int_y^\infty (z-y)\lambda e^{-\lambda z} \, dz$$

$$= hy + \frac{(h+p+r)}{\lambda} \exp(-\lambda y) - \frac{h}{\lambda}.$$

We are to find the numbers y^*, x^* such that, from (6.19),

$$Pr[Z \leqslant y^*] = 1 - \exp(-\lambda y^*) = (p+r-c)/(p+r+h)$$

and, from (6.20),

$$cx^* + hx^* + \frac{(h+p+r)}{\lambda} \exp(-\lambda x^*) = k + cy^* + hy^* + \frac{(h+p+r)}{\lambda} \exp(-\lambda y^*).$$

After a little further manipulation, it can be shown that

$$\exp[\lambda(y^* - x^*)] = 1 + \lambda(y^* - x^*) + \lambda k/(h+c),$$

which can be solved, in principle, for $y^* - x^*$.

Problem 8. If $k = \dfrac{1}{64}$, $c = \dfrac{1}{2}$, $h = 1$, $p = \dfrac{1}{4}$, $r = \dfrac{3}{4}$, and the demand is uniformly distributed over $(0,1)$ calculate x^* and y^* by repeating the analysis of section 6.4. ∎

6.5 ORDERING POLICY FOR TWO PERIODS

Here we envisage ordering at the start of the first period, settling up in some sense at the end of the first period and ordering again for the second period. The introduction of a second period brings a variety of complications in its train.

The main new point to settle is how to treat a shortage of units at the end of the first period. There are two main alternatives:

(a)   At the end of the first period, if demand has exceeded supply then we can allow the excess demands to remain unfulfilled. We say in this case that there is *no backlogging*.

(b)   If demand has exceeded supply, then a waiting list is kept. At the beginning of the second period, it is possible to purchase enough items for this waiting list and to set up the inventory for the second period. We say there is *backlogging*.

At the end of the second period, unsatisfied demand is usually lost under either scheme (a) or (b). If demand is less than supply, then the same holding charges may apply for both periods. In some schemes, a salvage value may be assigned to items finally unsold.

At first sight it would appear that the optimum policy for two periods should consist of repeating for the second period the optimum policy for the first period. This tempting solution is not in general correct. This is in part explained by the observation that $\min_{x}[A(x) + B(x)] \leqslant A(x) + B(x)$ for all x, *including the value of x which minimises $A(x)$ in isolation.*

There is yet another consideration arising from the fact that bills paid or income garnered at the end of the second period, being further removed in time, are not entirely equivalent to similar sums arising at the end of the first period.

Suppose a bill for an amount B has to be paid at the start of the second period. We can invest a smaller amount, A, at the beginning of this first period, which together with the interest earned over the first period, will equal B. Thus, if i is the interest rate per cent per period,

$$A + \frac{iA}{100} = B, \text{ giving } A = \alpha B,$$

where $\alpha = 1 / \left[1 + \dfrac{i}{100} \right].$

We say that A is the *present value* of B, or A is the **discounted value** of B. Similarly a revenue collected at the end of the second period is worth less than the same sum collected at the end of the first period. It is customary to *discount* all such delayed costs and revenues by a common factor α, where $0 < \alpha \leqslant 1$ and, strictly speaking, α depends on the duration of the period.

6.6 TWO PERIODS: NO BACKLOGGING

Suppose at the beginning of each period it is possible to purchase items at c each and these are delivered immediately. Revenue is r per item sold. Let $L(y)$ be the expected holding, shortage and lost sales cost for any period for which the initial inventory is y. Let $\ell_1(y_1, x_1)$ be the expected loss for both periods when at the start of the first period the store contains x_1 and is raised to y_1 while $\ell_2(y_2, x_2)$ is similarly the expected loss for the second period. Strictly we should write $\ell_2(y_2, X_2)$ to stress that the content of the store at the beginning of the second period is a random variable and depends on the demand in the first period. Indeed if the demands in the course of periods one and two are Z_1, Z_2 then $X_2 = y_1 - Z_1$. We shall assume that Z_1, Z_2 are independent, continuous random variables with a common p.d.f. For the second period we minimise $\ell_2(y_2, y_1 - Z_1)$ for $y_2 \geqslant y_1 - Z_1$ but the result, written $\ell_2(y_1 - Z_1)$, depends on $y_1 - Z_1$. So this minimum loss, after discounting, must be added to the loss for the first period, the sum averaged over the distribution of Z_1 and the result minimised for $y_1 \geqslant x_1$. That is to say, $\ell_1(x_1) = \min\limits_{y_1 \geqslant x_1} [\ell_1(y_1, x_1)]$

$$= \min_{y_1 \geqslant x_1} [c(y_1 - x_1) + L(y_1) - rE(Z_1) + \alpha \int_0^\infty \ell_2(y_1 - z_1) f(z_1) dz_1]. \quad (6.21)$$

(We subsequently omit capital letters for random variables where the context makes their presence obvious.)

One feature of equation (6.21) deserves special attention. If $z_1 \geqslant y_1$, the store runs out and *since there is no backlogging*, $\ell_2(y_1 - z_1) = \ell_2(0)$ for $z_1 \geqslant y_1$. Hence

$$\ell_1(x_1) = \min_{y_1 \geqslant x_1} [c(y_1 - x_1) + L(y_1) - rE(Z_1)$$

$$+ \alpha \left\{ \ell_2(0) \int_{y_1}^\infty f(z_1) dz_1 + \int_0^{y_1} \ell_2(y_1 - z_1) f(z_1) dz_1 \right\}]. \quad (6.22)$$

For one case, we know the form of $\ell_2(.)$ — namely when the holding and shortage costs are linear. We have, from equations (6.13), (6.14), for some y_2^*,

$$\begin{aligned} \ell_2(x_2) &= c(y_2^* - x_2) + L(y_2^*) - rE(Z_2) &&\text{for } 0 \leqslant x_2 < y_2^*. \\ \ell_2(x_2) &= L(x_2) - rE(Z_2) &&\text{for } y_2^* \leqslant x_2. \end{aligned} \right\} \quad (6.23)$$

In terms of the quantities appearing in (6.22),

$$\left. \begin{array}{lll} 0 \leqslant y_1 - z_1 < y_2^* & \text{implies} & z_1 > y_1 - y_2^* \text{ and } z_1 \leqslant y_1. \\ y_2^* \leqslant y_1 - z_1 & \text{implies} & z_1 \leqslant y_1 - y_2^*. \end{array} \right\} \qquad (6.24)$$

Hence, if we write (6.22) as

$$\ell_1(x_1) = -cx_1 + \min_{y_1 \geqslant x_1} G(y_1), \text{ where}$$

$$G(y_1) = cy_1 + L(y_1) - rE(Z_1) + \alpha \{ \ell_2(0) \int_{y_1}^{\infty} f(z_1)dz_1$$

$$+ \int_0^{y_1} \ell_2(y_1 - z_1)f(z_1)dz_1 \},$$

we must look at the minimisation of $G(y_1)$. We have

$$G'(y_1) = c + L'(y_1) + \alpha \{ -\ell_2(0)f(y_1) + \int_0^{y_1} \ell_2'(y_1 - z_1)f(z_1)dz_1$$

$$+ \ell_2(0)f(y_1) \}$$

$$= c + L'(y_1) + \alpha \int_0^{y_1} \ell_2'(y_1 - z_1)f(z_1)dz_1. \qquad (6.25)$$

For the second derivative we must be more careful. From (6.23), $\ell_2'(x_2) = -c$ for $0 \leqslant x_2 < y_2^*$, and $\ell_2'(x_2) = L'(x_2)$ for $y_2^* \leqslant x_2$. Incorporating these results in (6.25) we have, using (6.24),

$$G'(y_1) = c + L'(y_1) + \alpha \{ -c \int_{y_1 - y_2^*}^{y_1} f(z_1)dz_1$$

$$+ \int_0^{y_1 - y_2^*} L'(y_1 - z_1)f(z_1)dz_1 \} \qquad (6.26)$$

$$G''(y_1) = L''(y_1) + \alpha \{ -cf(y_1) + cf(y_1 - y_2^*) + \int_0^{y_1 - y_2^*} L''(y_1 - z_1)f(z_1)dz_1$$

$$+ L'(y_2^*)f(y_1 - y_2^*) \}$$

$$= L''(y_1) - \alpha cf(y_1) + \int_0^{y_1 - y_2^*} L''(y_1 - z_1)f(z_1)dz_1, \qquad (6.27)$$

since y_2^* satisfies $c + L'(y_2^*) = 0$.

Now we have struggled towards (6.27) with a view to seeing whether or not $G(y_1)$ is convex. Unfortunately, even if $L(.)$ is convex, the presence of the term $-\alpha cf(y_1)$ spoils the possibility of claiming that $G''(y_1) > 0$. However, in the case of linear holding and shortage costs,

$$L(y) = h \int_0^y (y-z)f(z)dz + (p+r) \int_y^\infty (z-y)f(z)dz$$

$$\Rightarrow L''(y) = (h+p+r)f(y).$$

Substituting in (6.27), we obtain

$$G''(y_1) = (h+p+r-\alpha c)f(y_1) + \int_0^{y_1-y_2^*}(h+p+r)f(y_1-z_1)f(z_1)dz_1 > 0,$$

since $(h+p+r-\alpha c) > (h+c-\alpha c) > 0$ when $0 < \alpha < 1$ and $c < p+r$. So in this case we do have a simple optimal policy. Let y_1^* satisfy $G'(y_1^*) = 0$ and y_2^* satisfy $c + L'(y_2^*) = 0$, then at the beginning of the first period:

if $x_1 < y_1^*$ order $y_1^* - x_1$ but if $x_1 \geqslant y_1^*$ order nothing.

At the beginning of the second period:

if $x_2 < y_2^*$ order $y_2^* - x_2$ but if $x_2 \geqslant y_2^*$ order nothing.

For this linear case, little change is involved when using a different holding and shortage loss for each period.

Example 8

Holding charge is two units per item in excess at the end of a period. Penalty charge is half a unit per item in deficit at the end of a period. Purchase cost is one unit per item ordered at the start of a period. Orders are delivered, immediately and no backlogging is allowed. Demands are independently and uniformly distributed over (0,16) per period. Discount rate for the second of two periods is 64/143. Revenue of one and a half units per unit sold.

The expected holding and shortage costs for a period with initial inventory y is, including lost revenue,

$$L(y) = 2 \int_0^y \frac{(y-z)}{16} dz + 2 \int_y^{16} \frac{(z-y)}{16} dz$$

$$= [y^2 + (16-y)^2]/16, \quad 0 \leqslant y \leqslant 16.$$

Hence $L'(y) = (y-8)/4.$

The number y_2^* must satisfy $c + L'(y_2^*) = 0$. That is $1 + (y_2^* - 8)/4 = 0$ or $y_2^* = 4$.

From equation (6.26) the number y_1^* must satisfy

$$0 = 1 + \frac{y_1^*}{4} - 2 + \frac{64}{143} \left\{ \frac{(y_1^*-4)}{16} - \frac{y_1^*}{16} + \int_0^{y_1^*-4} \frac{(y_1^*-z_1-8)}{64} dz_1 \right\}$$

$$= \frac{y_1^*}{4} - 1 + \frac{64}{143} \left\{ \frac{1}{4} - \left[\frac{(y_1^*-z_1-8)^2}{128} \right]_0^{y_1^*-4} \right\}$$

$$= \frac{y_1^*}{4} - 1 + \frac{64}{143}\left\{-\frac{1}{4} - \frac{1}{8} + \frac{(y_1^*-8)^2}{128}\right\}$$

$$= \frac{y_1^*}{4} - 1 + \frac{64}{143}\left\{\frac{1}{8} - \frac{y_1^*}{8} + \frac{(y_1^*)^2}{128}\right\}$$

$$= \frac{1}{572}(2y_1^* - 9)(y_1^* + 60), \text{ hence } y_1^* = 9/2.$$

If the store contained material rather than discrete items, the solution would serve. As it is, the choice between $y_1^* = 5$, $y_2^* = 4$ and $y_1^* = 4$, $y_2^* = 4$ can be decided by comparing the expected losses.

Problem 9. Show for Example 8 that $\ell_2(x) = 2 - x$ for $x < 4$ and $\ell_2(x) = (x^2 - 16x + 32)/8$ for $x \geqslant 4$. ∎

Example 9
Calculate the expected loss, for Example 8, for the policy $y_1^* = 5$, $y_2^* = 4$. We must be careful to distinguish the various possibilities for the contents of the store at the end of the first period. Suppose initially, that the store contains $x_1 < 5$: then the level must be raised to 5 at cost $5 - x_1$. Subsequently there are three cases to consider, determined by the demand z_1 during the first period.

(i) $5 \leqslant z_1 \leqslant 16$, the store is empty and the expected loss, over the demand for the second period, is 2 (see Problem 9). Taking account of the variation in z_1, the contribution to the second period is $\int_5^{16} (2/16)dz_1$.

For the first period there is a shortage penalty of $2 \int_5^{16} \frac{(z_1-5)}{16} dz_1$.

Hence, for this case, the total contribution is

$$2 \int_5^{16} \frac{(z_1-5)}{16} dz_1 + \alpha \int_5^{16} \frac{2}{16} dz_1.$$

(ii) $1 \leqslant z_1 < 5$, the store contains $5 - z_1$ at the end of the first period and is less than 4. Hence the level must be raised to 4, that is $z_1 - 1$ must be ordered. From Problem 9, the expected loss over the second period is $2 - (5 - z_1) = z_1 - 3$. The contribution to the second period is

$\int_1^5 [(z_1 - 3)/16] dz_1$. For the first period there is a holding penalty of

$2 \int_1^5 [(5 - z_1)/16] dz_1$. Hence for this case, the total contribution is

$$2 \int_1^5 \frac{(5-z_1)}{16} dz_1 + \alpha \left[\int_1^5 \frac{(z_1-3)}{16} dz_1\right].$$

(iii) $0 \leqslant z_1 < 1$, the store contains $5 - z_1$ at the end of the first period and is greater than 4. Hence nothing is ordered and, from Problem 9, the expected loss over the second period is $[(5-z_1)^2 - 16(5-z_1)+32]/8$. The total contribution to first and second periods is

$$2 \int_0^1 \frac{(5 - z_1)}{16} \, dz_1 + \alpha \left[\int_0^1 \frac{[(5-z_1)^2 - 16(5-z_1)+32]/8}{16} \, dz_1 \right]$$

The expected loss for this policy when $x_1 < 5$ is the sum of the total contributions from (i), (ii), (iii) together with an amount $5 - x_1 - 12$ which includes the initial purchase cost and the adjustment for incorporating lost revenue. This procedure saves computing the expected revenue for each case.

Problem 10. Complete Example 9 by finding the components of the expected loss for the policy $y_1^* = 5, y_2^* = 4$ when the store initially contains $x_1 > 5$. ∎

6.7 MULTIPLE PERIODS: NO BACKLOGGING

The method employed for two periods can be extended to deal with more than two periods; however, this leads to extremely tiresome computations. By making some further assumptions, we can obtain a policy that would be close to the optimum when the number of periods is large. The costs will be as for the two-period case. The ordering cost is c per unit quantity, but there is no set-up cost. Revenue is r per unit sold. Demands are independently and continuously distributed in each period. The holding and shortage penalty function is $L(.)$. Deliveries are immediate and there is a discount factor of α. We now assume that in the optimal policy for n periods, the level below which the inventory is not alllowed to fall, tends to a common value, y^*, for the start of each period. To implement a policy, we need to know the best choice of y^*.

Suppose the store is initially empty, then the expected loss for the first period is $cy^* + L(y^*) - rE(Z)$. At the start of the second period, the contents must be less than y^* and must be raised again to y^*. Hence we have a contribution $L(y^*)$ to the expected loss for the second period. The expected ordering cost is not so self-evident. If the demand, Z_1, in the first period, exceeds y^* then a quantity y^* must be purchased. If Z_1 is less than y^*, then the deficit, Z_1, must be ordered. Hence the expected ordering cost is

$$c \int_{y^*}^\infty y^* f(z_1) \, dz_1 + c \int_0^{y^*} z_1 f(z_1) \, dz_1$$

$$= cy^* [1 - F(y^*)] + c [y^* F(y^*) - \int_0^{y^*} F(z_1) \, dz_1]$$

$$= cy^* - c \int_0^{y^*} F(z_1) \, dz_1.$$

Hence the expected contribution to the total loss from the second period is, after discounting,

$$\alpha \left[cy^* - c \int_0^{y^*} F(z_1)dz_1 + L(y^*) - rE(Z_1) \right].$$

Similarly from the ith period $(i \geqslant 2)$ we have an expected loss

$$\alpha^{i-1} \left[cy^* - c \int_0^{y^*} F(z)dz + L(y^*) - rE(Z) \right].$$

Hence, in the limit, the total expected loss for this policy is

$$\ell(y^*,0) = cy^* + L(y^*) - rE(Z) + \sum_{i=2}^{\infty} \alpha^{i-1} \left[cy^* - c \int_0^{y^*} F(z)dz \right.$$
$$\left. + L(y^*) - rE(Z) \right]$$

$$= \left[cy^* + L(y^*) - \alpha c \int_0^{y^*} F(z)dz - rE(Z) \right] / (1-\alpha).$$

We minimise $\ell(y^*,0)$ with respect to y^* by equating $\ell'(y^*,0)$ to zero and obtain

$$c + L'(y^*) - \alpha c F(y^*) = 0. \tag{6.28}$$

If the initial stock level, x, is less than y^* then

$$\ell(y^*,x) = -cx + \ell(y^*,0).$$

We do not pursue the case when x exceeds y^*, for in the course of a long sequence of periods, the stock level must eventually fall below y^*.

Example 10
Linear holding and shortage costs.

$$L(y) = h \int_0^y (y-z)f(z)dz + (p+r) \int_y^{\infty} (z-y)f(z)dz,$$

hence $L'(y) = hF(y) - (p+r) [1 - F(y)]$.
    From (6.28),

$$c + hF(y^*) - (p+r) [1 - F(y^*)] - \alpha c F(y^*) = 0,$$

or

$$F(y^*) = (p+r-c)/(h+p+r-\alpha c).$$

In particular, when $h = 2, p = \frac{1}{2}, r = 3/2, c = 1, \alpha = 64/143$,

$$F(y^*) = 143/508.$$

Hence if Z is uniformly distributed over $(0,16)$, $y^* = (143/508)16$ which is close to the value of y_1^* found in Example 8 for the first of two periods.

Problem 11. Here, the result (6.28) is to be recovered by an extension of the method for two periods. Let $\ell_{1,n}(.)$ be the minimum expected loss from the beginning of the first period to the end of n periods, and $\ell_{2,n-1}(.)$ be the minimum expected loss from the beginning of the second period to the end of the remaining $n-1$ periods. With the same framework of costs as in section 6.7, if $\lim_{n\to\infty} \ell_{1,n}(x) = \lim_{n\to\infty} \ell_{2,n-1}(x) = \ell(x)$, show that this common limit is

$$= \min_{y \geqslant x} [c(y-x) + L(y) - rE(Z) + \alpha\{\ell(0) \int_y^\infty f(z)dz + \int_0^y \ell(y-z)f(z)dz\}].$$

Suppose we write $\ell(x) = -cx + \min_{y \geqslant x} G(y) - rE(Z)$. Assuming that $G(y)$ has a minimum at y^* and that this is attained at the only zero of $G'(y)$, show that

$$c + L'(y^*) - \alpha cF(y^*) = 0. \quad \blacksquare$$

6.8 TWO PERIODS: NO BACKLOGGING, FIXED ORDERING COST

We have already seen that the imposition of a fixed charge, k, for any positive order, in addition to the purchase price of c per unit quantity, complicates the computation of the optimum policy for one period. At this stage we restrict attention to the linear holding and penalty costs and allow ourselves to be guided by the results discovered for one period in section 6.4. These suggest that in the policy which has minimum expected loss, there are two pairs of numbers, $(x_1^*, y_1^*), (x_2^*, y_2^*)$ such that

(a)   if the contents at the start of the first period, x_1 do not exceed x_1^* order $y_1^* - x_1$, otherwise ordering nothing.

(b)   if the contents at the start of the second period, x_2, do not exceed x_2^* order $y_2^* - x_2$, otherwise order nothing.

For the second period, we know how to find y_2^*, x_2^*, they satisfy

$$c + L'(y_2^*) = 0,$$
$$cx_2^* + L(x_2^*) = k + cy_2^* + L(y_2^*). \tag{6.29}$$

We seek an analogous pair of equations for $x_1^* y_1^*$. With the same pattern as (6.21), but omitting the constant $-rE(Z_1)$,

$$\ell_1(x_1) = \min_{y_1 \geqslant x_1} [k + c(y_1 - x_1) + L(y_1) + \alpha \int_0^\infty \ell_2(y_1 - z_1)f(z_1)dz_1], \tag{6.30}$$

where $k > 0$ if $y_1 > x_1$ and $k = 0$ if $y_1 = x_1$. The condition on k is undeniably clumsy to incorporate in the notation. The reader should bear in mind the equivalent formulation

$$\ell_1(x_1) = \min \begin{cases} \min_{y_1 > x_1} \left[k + c(y_1 - x_1) + L(y_1) + \alpha \int_0^\infty \ell_2(y_1 - z_1) f(z_1) dz_1 \right] & (6.31) \\ \\ L(x_1) + \alpha \int_0^\infty \ell_2(x_1 - z_1) f(z_1) dz_1 & (6.32) \end{cases}$$

$$= \min \begin{cases} -cx_1 + k + \min_{y_1 > x_1} G(y_1) & (6.33) \\ \\ -cx_1 \qquad + \qquad G(x_1). & (6.34) \end{cases}$$

Suppose y_1^* is the unconstrained minimum of $G(y_1)$. If y_1^* satisfies $G'(y_1^*) = 0$, then bearing in mind that there is no backlogging we have, from (6.25),

$$c + L'(y_1^*) + \alpha \int_0^{y_1^*} \ell_2'(y_1^* - z_1) f(z_1) dz_1 = 0. \qquad (6.35)$$

As remarked in 6.4, there may be values $x_1 < y_1^*$, such that $G(x_1) < k + G(y_1^*)$. The other number x_1^* that we seek satisfies

$$G(x_1^*) = k + G(y_1^*). \qquad (6.36)$$

We have given a somewhat superficial account of a calculation which deserves closer analysis. This it will receive in a later (optional) section. We give first an illustrative example.

Example 11
If $k = 1/64$, $c = 1/2$, $h = 1$, $p = 1/4$, $r = 3/4$ and demand is uniformly distributed over $(0,1)$ in each period, find equations to determine the optimal policy for two periods when the discount factor is α.

In Problem 8 we have already found the optimal solution for the second period. We repeat the results. $x_2^* = \frac{1}{8}$, $y_2^* = \frac{1}{4}$, $L(x) = x^2 - x + \frac{1}{2}$, $\ell_2(x) = x^2 - x + \frac{1}{8}$ $\left(x \geq \frac{1}{8} \right)$, $\ell_2(x) = (5 - 32x)/64$ $\left(x < \frac{1}{8} \right)$. Hence $\ell_2'(x) = 2x - 1$ $(x \geq 1/8)$ and $\ell_2'(x) = -1/2$, $x < 1/8$. Thus in the integral in (6.35) we need to split the range.

$$\int_0^{y_1 - 1/8} [2(y_1 - z_1) - 1] dz_1 + \int_{y_1 - 1/8}^{y_1} -\frac{1}{2} dz_1 = y_1^2 - y_1 + \frac{3}{64}.$$

Hence $G'(y_1) = 2y_1 - \frac{1}{2} + \alpha\left(y_1^2 - y_1 + \frac{3}{64}\right)$. Now $G'(y_1)$ has at most two real zeros. Since $G'\left(\frac{1}{4}\right) < 0$ and $G'(1) > 0$, one lies between $\frac{1}{4}$ and 1 while the other is negative and is irrelevant. In fact, $G(y_1)$ is convex for $0 < y_1 < 1$. In particular if $\alpha = 96/101$, the reader should confirm that $y_1^* = 1/3$. To find x_1^*, use (6.36).

Problem 12. In Example 11, we assumed that $y_1^* > 1/8$. By considering $G'(y_1)$ for $0 < y_1 < 1/8$, show that the assumption is justified. ■

*6.9   FURTHER DISCUSSION OF THE EFFECT OF A FIXED ORDERING COST

For the optimal solution to have the form suggested in (a), (b) in section 6.8, everything turns on the behaviour of $G(.)$. Problem 13 demonstrates that, even in the linear case, the convexity of $G(.)$ is not guaranteed. This is to be contrasted with section 6.6 where there was no fixed ordering charge and, in the linear case, $G(.)$ was convex.

Problem 13. If, in the framework of section 6.8,

$$G'(y_1) = c + L'(y_1) + \alpha \int_0^{y_1} \ell_2'(y_1 - z_1)f(z_1)\mathrm{d}z_1,$$

show that $G''(y_1)$ is not necessarily positive, even when all the costs are linear. ■

The existence of an (s,S) policy for both periods is closely linked with the behaviour of $G'(.)$. For $G(.)$ to have one turning value, which corresponds to a minimum, $G'(.)$ must show one change of sign – from negative to positive. Now

$$G'(y_1) = c + L'(y_1) + \alpha \int_0^{y_1} \ell_2'(y_1 - z_1)f(z_1)\mathrm{d}z_1. \qquad (6.37)$$

In the case of linear costs, $L'(y_1) = hF(y_1) - (p+r)[1-F(y_1)]$: hence we may write (6.37) as

$$G'(y_1) = \int_0^{y_1} [c+h+\alpha\,\ell_2'(y_1-z_1)]\, f(z_1)\mathrm{d}z_1 + (c-p-r)\int_{y_1}^{\infty} f(z_1)\mathrm{d}z_1 \qquad (6.38)$$

Now $G'(0) = c-p-r < 0$ if $c < p+r$ (our usual requirement). We can make $\int_{y_1}^{\infty} f(z_1)\mathrm{d}z_1$ as small as we please by choosing y_1 to be sufficiently large. Hence $G'(y_1)$ will eventually be positive if $c+h+\alpha\,\ell_2'(y_1-z_1)$ is positive. Now $\ell_2'(x_2) = -c$ for $x_2 < x_2^*$ and $\ell_2'(x_2) = L'(x_2)$ for $x \geqslant x_2^*$. Hence $c+h+\alpha\,\ell_2'(x_2)$

is trivially positive for $x_2 < x_2^*$. It can readily be shown that for $x_2 \geqslant x_2^*$, $c+h+\alpha \, \ell_2'(x_2)$ is

$$[c+h(1+\alpha)] \int_0^{x_2} f(z_2)\mathrm{d}z_2 + [c+h-\alpha(p+r)] \int_{x_2}^{\infty} f(z_2)\mathrm{d}z_2 \qquad (6.39)$$

and will certainly be positive provided $(c+h) > \alpha(p+r)$. The two conditions, $c < p+r$, $c+h > \alpha(p+r)$, ensure that $G'(y_1)$ has at least one change of sign. Whether or not there is just one change of sign depends on the properties of $f(.)$. It is sufficient to remark here that many of the standard distributions, (for instance gamma, normal, Poisson), have p.d.f.s which give one change of sign in $G'(.)$ for the linear case expressed in equation (6.38). For a full discussion of the technical points, the reader should consult Karlin[3].

6.10 TWO PERIODS WITH BACKLOGGING

Suppose unsatisfied demands during the first period are backlogged. Then at the beginning of the second period we must order enough items to discharge the backlog and to optimise for the second period. If there is a cost of c per unit quantity ordered, but no fixed charge for placing an order, then $\ell_1(x_1)$, the minimum expected loss from the beginning of the first period, when there is an amount x_1 on hand before placing an order for an amount $y_1 - x_1$, satisfies

$$\ell_1(x_1) = \min_{y_1 \geqslant x_1} \left[c(y_1 - x_1) + L_1(y_1) - rE(Z_1) + \alpha \int_0^{\infty} \ell_2(y_1 - z_1)f(z_1)\mathrm{d}z_1 \right].$$
$$(6.40)$$

Equation (6.40) differs from Equation (6.22) in two important respects. Since backlogging is allowed, the argument of $\ell_2(.)$ may be negative. Also we have written $L_1(y_1)$ for the expected holding and penalty cost for an intitial inventory of y_1, since additional revenue has to be incorporated from the sale of backlogged items. Most of our attention will be directed to the case of linear holding and penalty costs and when there is a revenue of r per unit quantity sold.

If backlogging is allowed, then at the end of the first period we can place an order sufficient not only to meet previously unsatisfied demand but also to reach an optimum level for the second period. At first sight, it might seem that we should thus begin the first period with little or no stock since we can cover the orders later and avoid holding charges. In practice, the infliction of shortage penalties may make this undesirable.

We notice a slight dilemma over the liquidation of the backlog. The additional revenue is only recovered during the second period, so it must be discounted. Thus the expected holding and penalty costs incurred at the end of the first period, in the linear case, are

$$h \int_0^{y_1} (y_1-z_1)f(z_1)dz_1 + p \int_{y_1}^{\infty} (z_1-y_1)f(z_1)dz_1 .$$

On the other hand, the expected revenue is

$$r \int_0^{y_1} z_1 f(z_1)dz_1 + ry_1 \int_{y_1}^{\infty} f(z_1)dz_1 + \alpha r \int_{y_1}^{\infty} (z_1-y_1)f(z_1)dz_1 ,$$

$$= rE(Z_1) + (\alpha r - r) \int_{y_1}^{\infty} (z_1-y_1)f(z_1)dz_1 .$$

Hence the expected loss for the period is $L_1(y_1) - rE(Z_1)$

$$= h \int_0^{y_1} (y_1-z_1)f(z_1)dz_1 + (p+r-\alpha r) \int_{y_1}^{\infty} (z_1-y_1)f(z_1)dz_1 - rE(Z_1). \quad (6.41)$$

The total 'penalty' per unit quantity in deficit is now $p+r-\alpha r$, rather than $p+r$ as in the lost sales case. If $\alpha = 1$, there is no additional penalty for lost revenue. Now, for the second period, we know the optimal rule. There is a number y_2^*, such that if the amount, x_2, left at the end of the first period is less than y_2^* we raise the inventory level to y_2^* — otherwise we order nothing. Now $x_2 = y_1-z_1$, where z_1 is the demand in the first period, and may be negative. If the initial inventory y_1 is less than y_2^* the minimum expected loss for the second period is for all z_1,

$$\ell_2(x_2) = \ell_2(y_1-z_1) = c(y_2^*-y_1+z_1)+L_2(y_2^*)-rE(Z_2). \quad (6.42)$$

On the other hand, if $y_1 > y_2^*$, then when $z_1 > y_1 - y_2^*$

$$\ell_2(y_1-z_1) = c(y_2^*-y_1+z_1) + L_2(y_2^*) - rE(Z_2) \quad (6.43)$$

but when $z \leqslant y_1 - y_2^*$, we have

$$\ell_2(y_1-z_1) = L_2(y_1-z_1) - rE(Z_2). \quad (6.44)$$

These results must be born in mind when applying (6.40).

Example 12

There is no fixed charge; unit volume costs one unit and orders are delivered immediately. There is a shortage charge cost of one unit and a holding charge of two units per unit volume, both levied on the inventory level at the end of each period. Backlogging takes place at the end of the first period but not at the end of the second. Revenue is two units per unit volume sold. The demands for each period are independent and uniformly distributed between zero and fifteen. The discount factor is α. Here we do not have the same expected costs and returns for the two periods. If y_2 is the initial inventory for the second period, then the expected holding and shortage cost for the second period is

$$L_2(y_2) = h \int_0^{y_2} (y_2 - z_2) f(z_2) dz_2 + (p+r) \int_{y_2}^{\infty} (z_2 - y_2) f(z_2) dz_2$$

$$= 2 \int_0^{y_2} \frac{(y_2 - z_2)}{15} dz_2 + 3 \int_{y_2}^{15} \frac{(z_2 - y_2)}{15} dz_2$$

$$= \frac{y_2^2}{15} + \frac{(15 - y_2)^2}{10}.$$

The total expected loss for the second period is

$$\ell_2(y_2, x_2) = c(y_2 - x_2) + L_2(y_2) - rE(Z_2)$$

$$= y_2 - x_2 + [2y_2^2 + 3(15 - y_2)^2]/30 - 15,$$

where x_2 is the stock at the end of the first period. Now $y_2 + [2y_2^2 + 3(15 - y_2)^2]/30$ has a minimum at $y_2^* = 6$. Hence the optimal policy is to raise the inventory to 6 if $x_2 < 6$ and otherwise order nothing. Hence the minimum expected loss for the second period satisfies

$$\ell_2(x_2) = \min_{y_2 \geqslant x_2} [\ell_2(y_2, x_2)],$$

$$= \begin{cases} (6 - x_2) + L_2(6) - 15 & \text{for } x_2 < 6 \\ L_2(x_2) - 15 & \text{for } x_2 \geqslant 6 \end{cases}$$

$$= \begin{cases} \dfrac{3}{2} - x_2 & \text{for } x_2 < 6 & (6.45) \\[2mm] \dfrac{x_2^2}{6} - 3x_2 + \dfrac{15}{2} & \text{for } x_2 \geqslant 6. & (6.46) \end{cases}$$

We are now ready to tackle two periods. From (6.41), the holding and shortage function for the first period is

$$L_1(y_1) = 2 \int_0^{y_1} \frac{(y_1 - z_1)}{15} dz_1 + (1 + 2 - 2\alpha) \int_{y_1}^{15} \frac{(z_1 - y_1)}{15} dz_1$$

$$= \frac{y_1^2}{15} + \frac{(3 - 2\alpha)(15 - y_1)^2}{30}.$$

Now suppose that $y_1 < 6$: then perforce $y_1 - z_1 < 6$ and for the second period (6.45) is appropriate. Hence (6.40), after substituting, becomes

$$\min_{y_1 \geqslant x_1} \left\{ y_1 - x_1 + \frac{y_1^2}{15} + \frac{(3 - 2\alpha)(15 - y_1)^2}{30} - 15 + \alpha \int_0^{15} \frac{[(3/2) - y_1 + z_1]}{15} dz_1 \right\},$$

$$(6.47)$$

$$= -x_1 + \min_{y_1 \geqslant x_1} [G(y_1)].$$

The derivative of $G(y_1)$ is

$$1 + \frac{2y_1}{15} - \frac{(3-2\alpha)(15-y_1)}{15} - \alpha$$

$$= -2 + \alpha + y_1(5-2\alpha)/15.$$

Clearly $G(y_1)$ has a minimum at $y_1^* = 15(2-\alpha)/(5-2\alpha)$, and for all α, $5 \leqslant y_1^* \leqslant 6$. (In fact when $\alpha = 0$, the recoverable revenue from backlogged sales becomes worthless!) So long as $0 \leqslant y_1 < 6$, $G(y_1)$ is a quadratic in y_1. Problem 14 will confirm that values of $y_1 > 6$ need not be entertained and, in sum, that the optimum solution is (s,S) for both periods: viz.

If the initial contents are less than y_1^* then raise the inventory to $y_1^* = 15(2-\alpha)/(5-2\alpha)$.

If the contents at the end of the first period are less than y_2^* then raise the inventory to $y_2^* = 6$.

In particular when $\alpha = 1$, $y_1^* = 5$ and we proceed to compute $\ell_1(x_1)$ for $0 \leqslant x_1 < 5$. Here we must raise the level to 5 at the beginning of the first period and subsequently to 6 at the beginning of the second period. Hence, from (6.47),

$$\ell_1(x_1) = 5-x_1 + L_1(5) - 15 + 1 \int_0^{15} \frac{[(3/2-x_1+z_1]}{15} \, dz_1$$

$$= 5 - x_1 + 5 - 15 + \frac{3}{2} - x_1 + \frac{15}{2} = 4 - 2x_1.$$

Problem 14. For Example 12, when $\alpha = 1$, show that if

$$\ell_1(x_1) = -x_1 + \min_{y_1 \geqslant x_1} [G(y_1)], \text{ then } G'(y_1) > 0 \text{ for } y_1 > 6. \quad \blacksquare$$

Problem 15. For Example 12, when $\alpha = 1$ calculate $\ell_1(x_1)$ for $6 \leqslant x_1 \leqslant 15$. $\quad \blacksquare$

Problem 16. For Example 12 show that it is not worth ordering if the initial inventory level x_1, is such that $15 \leqslant x_1 < 21$, for any discount factor α. $\quad \blacksquare$

Problem 17. Suppose that the store is initially empty. If the number of periods is allowed to increase then the effect of no backlogging for the final period wears off. Show that if the inventory is to be raised to a common value y^* at the start of each period, then the best choice of y^* for this policy satisfies

$$c(1-\alpha) + L_1'(y^*) = 0.$$

(Hint: Find the expected loss for each period and recall that the loss for the ith period must be discounted by α^{i-1}.) $\quad \blacksquare$

*6.11 TWO PERIODS WITH BACKLOGGING, FIXED ORDERING COST

We shall not pursue this topic in any detail, since many of the issues involved have already been discussed in section 6.8. We proceed at once to an illustrative example.

Example 13
To obtain the optimal policy for Example 12 when an additional fixed cost of $k = 3/2$ is incurred for ordering any positive amount, and $\alpha = 1$.

In Example 12, we have sufficient information to find readily the optimum policy for the second period. We have

$$\ell_2(x_2) = \min \begin{cases} \min_{y_2 > x_2} \; [3/2 + (y_2 - x_2) + L_2(y_2) - 15]. \\ \\ L_2(x_2) - 15. \end{cases}$$

Now $L_2(y_2) = \dfrac{y_2^2}{15} + \dfrac{(15-y_2)^2}{10}$, from Example 12, hence

$$\ell_2(x_2) = \min \begin{cases} 3/2 - 15 - x_2 + \min_{y_2 > x_2} \; [y_2 + \{2y_2^2 + 3(15-y_2)^2\}/30]. \quad (6.48) \\ \\ -15 + [2x_2^2 + 3(15-x_2)^2]/30. \quad (6.49) \end{cases}$$

Now if $x_2 < 6$, $y_2 + L_2(y_2)$ has a minimum at $y_2^* = 6$, when (6.48) assumes the value $3/2 - 15 - x_2 + 6 + 21/2 = 3 - x_2$. We now seek those $x_2 (<6)$, such that

$$L_2(x_2) - 15 < 3 - x_2, \quad \text{or after re-arranging terms,}$$

$$(x_2 - 3)(x_2 - 9) < 0.$$

That is to say, $x_2^* = 3$. Hence, at the end of the first period, if the contents are below 3, raise the inventory to 6. Otherwise order nothing.

$$\ell_2(x_2) = 3 - x_2, \quad \text{for } x_2 < 3. \quad (6.50)$$

$$\ell_2(x_2) = \frac{x_2^2}{6} - 3x_2 + \frac{15}{2}, \quad \text{for } x_2 \geqslant 3. \quad (6.51)$$

We are now in a position to consider two periods. In the usual notation,

$$\ell_1(x_1) = \min \begin{cases} \min_{y_1 > x_1} \left[3/2 + (y_1 - x_1) + L_1(y_1) - 15 + \int_0^{15} \frac{\ell_2(y_1 - z_1)}{15} \, dz_1 \right] \\ \\ \hspace{10em} (6.52) \\ \\ \hspace{3em} L_1(x_1) - 15 + \int_0^{15} \frac{\ell_2(x_1 - z_1)}{15} \, dz_1 \\ \\ \hspace{12em} (6.53) \end{cases}$$

in which, from Example 12, $L_1(y_1) = \dfrac{y_1^2}{15} + \dfrac{(15-y_1)^2}{30}$.

It is readily shown that we may restrict attention to the case $y_1 > 3$. Then using (6.50) and (6.51),

$$\int_0^{15} \frac{\ell_2(y_1-z_1)}{15} dz_1 = \int_{y_1-3}^{15} \frac{(3-y_1+z_1)}{15} dz_1 + \int_0^{y_1-3} \left[\frac{(y_1-z_1)^2}{6} \right.$$

$$\left. - 3(y_1-z_1) + \frac{15}{2} \right] \frac{1}{15} dz_1. \qquad (6.54)$$

Hence the derivative of $y_1 + L_1(y_1) + \displaystyle\int_0^{15} \frac{\ell_2(y_1-z_1)}{15} dz_1$ is $\dfrac{1}{15}\left(\dfrac{y_1^2}{6} + y_1 - \dfrac{21}{2}\right)$ which is negative at $y_1 = 3$ and positive at $y_1 = 6$. There is a minimum at approximately $y_1^* = 11/2$, hence (6.52), for $x_1 < 11/2$ is

$$3/2 + (11/2 - x_1) + L_1(11/2) - 15 + \int_0^{15} \frac{\ell_2(11/2-z_1)}{15} dz_1. \qquad (6.55)$$

But of course there may be values of x_1 such that (6.53) is less than (6.55). A side calculation will confirm that we only need consider values of $x_1 < 3$, as we would expect since $x_2^* = 3$. Using (6.50), if $x_1 < 3$, then (6.53) is $3 - 2x_1 + x_1^2/10$. After some tedious arithmetic, which the reader may omit, we find that (6.55) is approximately $1.84 - x_1$. Hence it is still not worth ordering if

$$\frac{x_1^2}{10} - 2x_1 + 3 > 1.84 - x_1,$$

whence easily, $x_1 > 1.3$. The best policy is thus

If the initial contents at the start of the first period do not exceed 1.3, raise to 5.5, otherwise order nothing.
If the contents at the end of the first period do not exceed 3, raise to 6, otherwise order nothing.

Problem 18. For Example 13 show that (6.53) is $\dfrac{x_1^2}{10} - 2x_1 + 3$, when $x_1 < 3$. Evaluate (6.55). ∎

The general result for two periods, is, omitting $-rE(Z_1)$,

$$\ell_1(x_1) = \min \begin{cases} \displaystyle\min_{y_1 > x_1} \left[k + c(y_1-x_1) + L_1(y_1) + \alpha \int_0^\infty \ell_2(y_1-z_1) f(z_1) dz_1 \right] \\[4mm] \\ L_1(x_1) + \alpha \displaystyle\int_0^\infty \ell_2(x_1-z_1) f(z_1) dz_1, \end{cases}$$

$$= \min \begin{cases} -cx_1 + \min_{y_1 > x_1} [k + G(y_1)] & \text{(6.56)} \\ \\ -cx_1 + G(x_1). & \text{(6.57)} \end{cases}$$

If $G(y_1)$, as defined in (6.56), has only one change of sign and that at a minimum y_1^*, the procedure is straightforward. We locate x_1^*, such that, from (6.57), $G(x_1^*) = G(y_1^*) + k$ and then the policy for the first period is (x_1^*, y_1^*) as in Example 11. One change of sign limits the choice of demand distribution. We can in fact tolerate more than one minimum, providing that those which come after the first do not induce any ordering.

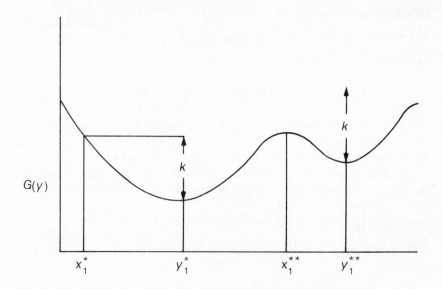

Figure 6.3

In Fig. 6.3, the first minimum of $G(y)$ induces raising the inventory to y_1^* if initially there are fewer than x_1^* items on hand. At the next minimum, however, $k + G(y_1^{**}) > G(x_1^{**})$, so that we are not forced to order if the number of items on hand lies between y_1^* and y_1^{**}. For the case of linear holding, shortage and revenue this property will hold when backlogging is allowed. For a full discussion the reader should consult Ross[4].

6.12 DISCRETE DEMAND

We have laid particular emphasis on the use of a continuous distribution for demand since such an assumption facilitates the solution of optimisation problems. The use of a discrete distribution is not so daunting when the inventory policy has been stipulated.

Example 14
The capacity of a store is two items. At the end of each week, if the store is empty, an order for two items is made, otherwise no order is made. Deliveries of replacements take place during the week-end and hence are effectively immediate. The probability that i items are demanded is $\phi_i(i = 0,1,2, \ldots)$ and demands in excess of stock are lost. Consider the number of items in stock *at the end of the week*. If p_{ij} is the conditional probability that the store contains j items at the end of any week *given* that the store contained i items at the end of the previous week, then we may easily check the values of the transition probabilities displayed in Table 6.1. Note that, if the store was empty before the last order, it had to be made up to two and will be empty again before the following order time if at *least* two items are demanded. The unconditional probability, $p_j^{(n)}$, that the store contains j items at the end of the nth week must satisfy

$$p_j^{(n)} = \sum_{i=0}^{2} p_i^{(n-1)} p_{ij}, \quad j = 0,1,2. \tag{6.58}$$

The equations (6.58) can be solved for $n = 1,2, \ldots$ but we are mainly interested in the limiting values $\pi_j = \lim_{n \to \infty} p_j^{(n)}$.

Table 6.1. Contents at end of a week

	i/j	0	1	2
	0	$1-\phi_0-\phi_1$	ϕ_1	ϕ_0
Contents at end	1	$1-\phi_0$	ϕ_0	0
of previous week	2	$1-\phi_0-\phi_1$	ϕ_1	ϕ_0

Problem 19. By considering the limits of both sides of (6.58) show that

$$\pi_j = \sum_{i=0}^{2} \pi_i p_{ij},$$

and verify that

$$\pi_0 = (1-\phi_0)^2/(1-\phi_0+\phi_1), \; \pi_1 = \phi_1/(1-\phi_0+\phi_1), \; \pi_2 = \phi_0(1-\phi_0)/(1-\phi_0+\phi_1). \; \blacksquare$$

Problem 20. In Example 14 consider the inventory level immediately *after* a delivery. Show that the limiting probabilities ρ_1, ρ_2 that the store contains 1,2 items are $\phi_1/(1-\phi_0+\phi_1)$, $(1-\phi_0)/(1-\phi_0+\phi_1)$ respectively. Check that $\rho_2 = \pi_2+\pi_0$ and $\rho_1 = \pi_1$. $\blacksquare$

Our main interest in such limiting probabilities is to obtain an approximate assessment of the cost of such a policy as outlined in Example 14. For instance, suppose a replacement item costs c and that there is a fixed charge k for any positive order. Let h be the holding charge per item remaining at the end of the week, p the penalty for each lost sale and r the revenue for each actual sale. We suppose that the costs for a period are composed of the re-ordering costs at its commencement and the holding, penalty and revenue which apply, on average, at the end of such a period. Suppose at the end of the week that the store is empty. Then two items must be purchased at cost $k + 2c$. At the end of the next week, there will be a holding charge if either both items are still there, with probability ϕ_0, or one item remains, with probability ϕ_1. The expected holding charge is thus $2h\phi_0 + h\phi_1$. There is a shortage penalty if there are at *least* three demands. Hence the expected shortage penalty is $p \sum_{j=3}^{\infty} (j-2)\phi_j$. Since both items are sold if the demand is at *least* two, the expected revenue is $r\phi_1 + 2r(1-\phi_0-\phi_1)$. The reader should verify Table 6.2.

Table 6.2

Inventory level	Holding charge	Shortage charge	Revenue
0	$2h\phi_0+h\phi_1$	$p \sum_{3}^{\infty} (j-2)\phi_j$	$r\phi_1+2r(1-\phi_0-\phi_1)$
1	$h\phi_0$	$p \sum_{2}^{\infty} (j-1)\phi_j$	$r(1-\phi_0)$
2	$2h\phi_0+h\phi_1$	$p \sum_{3}^{\infty} (j-2)\phi_j$	$r\phi_1+2r(1-\phi_0-\phi_1)$

Problem 21. From equation (6.58) and Table 6.1, show that

$$p_1^{(n)} = p_1^{(n-1)} (\phi_0-\phi_1) + \phi_1 .$$

Hence calculate $p_1^{(n)}$ and $\lim_{n \to \infty} [p_1^{(n)}]$.   ∎

6.13 DELIVERY LAG: FIXED DURATION

For the most part we have assumed that there was no delay between placing and receiving a re-stocking order. Such a delay is called a **delivery lag** and its duration is the **lead time**. We illustrate the effect of delivery lags through some particular examples. We start with a fixed lead time.

Re-orders can be made at regular intervals of time which are equal to the lead time. Re-orders are thus made immediately after a delivery except for the first. The policy shall be to re-order the number of items sold since the last order. The first re-order will be for the number of items sold from a full store. Backlogging is not permitted. It is convenient to consider the inventory level immediately after a delivery. Suppose customer demands are for single items and the probability of j customers in a lead time is ϕ_j. Let $p_r^{(n)}$ be the probability that the store contains r items immediately after the nth delivery. We obtain a recurrence relation in terms of the previous delivery ($n \geqslant 2$). For there will be $r(1 \leqslant r \leqslant S)$ in the store of capacity S after the nth delivery if either:

(a)   there were $S-i$ items in the store after delivery $n-1$, an order for $i(\leqslant r-1)$ items was sent out and just $S-r$ item demands made before this order is delivered. Note that $(S-i) - (S-r) + i = r$; or

(b)   there were $S-r$ items in the store after delivery $n-1$, an order for r was sent out and $(S-r)$ or *more* demands made before it is delivered.

$$p_r^{(n)} = \sum_{i=0}^{r-1} p_{S-i}^{(n-1)} \phi_{S-r} + p_{S-r}^{(n-1)} \sum_{j=S-r}^{\infty} \phi_j, \quad 1 \leqslant r \leqslant S. \qquad (6.59)$$

The case $r = 0$ is slightly different, for the store is empty if and only if no order was placed at the last order time and the full capacity of S items was sold by the current order point.

$$p_0^{(n)} = p_S^{(n-1)} \sum_{j=S}^{\infty} \phi_j. \qquad (6.60)$$

For the first delivery, it is easily checked that (6.59) and (6.60) hold for $n = 1$ if the store is initially full $[p_{S-i}^{(0)} = 0, i \neq 0]$.

Problem 22. Show that the expected number of sales before the first delivery, when the store is initially full, is

$$\sum_{j=1}^{S-1} j\phi_j + S \sum_{j=S}^{\infty} \phi_j. \quad \blacksquare$$

Problem 23. If $t_0, 2t_0, \ldots$ are the re-order times and demands have a Poisson distribution with mean λ per unit time, show that

$$p_S^{(n)} = \exp(-\lambda t_0) + [1 - \exp(-\lambda t_0)] \, p_0^{(n-1)}. \quad \blacksquare$$

6.14 DELIVERY LAG: RANDOM DURATION

To study the effect of a lead time which is variable, we place a severe restriction on its distribution. We suppose that the times between customer demands, which are for one item, have independent exponential distributions with common parameter λ. As soon as an item is sold, a replacement is ordered, the delivery time for which has, independently of all other events, another exponential distribution with parameter μ. No backlogging is allowed, so that unsatisfied demand represents lost sales. The store is initially full with capacity S. Let $p_n(t)$ be the probability that there are n items in the store, hence $S-n$ orders outstanding, at time t. Then there will be n $(0 < n < S)$ on hand at time $t + \delta t$, if

(a) there were $n-1$ at time t, no customer arrived and one of the outstanding $S-n+1$ orders arrives in time $(t, t+\delta t)$, this compound event having probability

$$p_{n-1}(t) \cdot (1-\lambda\delta t)\binom{S-n+1}{1}\mu\delta t(1-\mu\delta t)^{S-n} + o(\delta t)$$

$$= (S-n+1)\mu p_{n-1}(t)\delta t + o(\delta t);$$

or

(b) there were $n+1$ at time t, one customer arrived and none of the then out-standing $S-n-1$ orders arrived in time $(t, t+\delta t)$, this compound event having probability

$$p_{n+1}(t)\lambda\delta t(1-\mu\delta t)^{S-n-1} + o(\delta t)$$

$$= \lambda p_{n+1}(t)\delta t + o(\delta t);$$

or

(c) there were n at time t, no customer arrived and none of the $S-n$ orders then outstanding arrives, this compound event having probability

$$p_n(t) (1-\lambda\delta t) (1-\mu\delta t)^{S-n} + o(\delta t)$$

$$= p_n(t) \left\{ 1 - [\lambda+(S-n)\mu] \delta t \right\} + o(\delta t).$$

Collecting the terms in (a), (b), (c), and equating to $p_n(t+\delta t)$, we derive in the usual way the equation

$$\frac{dp_n(t)}{dt} = - [(S-n)\mu+\lambda]p_n(t) + \lambda p_{n+1}(t) + (S-n+1)\mu p_{n-1}(t). \qquad (6.61)$$

Also

$$\frac{dp_S(t)}{dt} = - \lambda p_S(t) + \mu p_{S-1}(t),$$

which is the same as equation (6.61) with $n = S$, since $p_{S+1}(t) \equiv 0$. Finally,

$$\frac{dp_0(t)}{dt} = - S\mu p_0(t) + \lambda p_1(t),$$

which is *not* the same as (6.61) when $n = 0$, even after setting $p_{-1}(t) \equiv 0$, a circumstance arising from inability to satisfy demand when the store is empty.

A necessary condition for a limiting distribution is $\lambda < \mu$, and if such a distribution exists, the limiting probabilities p_r must satisfy

$$\lambda(p_r - p_{r+1}) = \mu \{[S-(r-1)]p_{r-1} - (S-r)p_r\}, \quad 1 \leqslant r \leqslant S, \tag{6.62}$$

and

$$\lambda p_1 = S\mu p_0. \tag{6.63}$$

Summing (6.62) from $r = 1$ to n,

$$\lambda(p_1 - p_{n+1}) = [Sp_0 - (S-n)p_n]\mu, \tag{6.64}$$

and, using (6.63),

$$(S-n)\mu p_n = \lambda p_{n+1}, \quad 0 \leqslant n \leqslant S. \tag{6.65}$$

Together with the normalising condition $\sum_0^S p_r = 1$, equation (6.65) can be used to find p_n. We may also find the expected number of items in the store. Summing (6.65) from 0 to S

$$\mu \sum_{n=0}^S (S-n)p_n = \lambda \sum_{n=0}^S p_{n+1},$$

$$\mu S - \mu E(N) = \lambda(1-p_0)$$

and

$$E(N) = S - \frac{\lambda}{\mu}(1-p_0) < S.$$

Ordering in Batches

The opposite extreme re-ordering policy is to re-order S items as soon as the store is empty. Again, unsatisfied demand results in lost sales, but such a policy saves on delivery charges. By the same method as that employed above, we obtain the equations

$$\frac{dp_n(t)}{dt} = \lambda [p_{n+1}(t) - p_n(t)], 0 < n < S, \text{ there being nothing on order.}$$

$$\frac{dp_S(t)}{dt} = -\lambda p_S(t) + \mu p_0(t), \text{ for when empty a re-order is made; and}$$

$$\frac{dp_0(t)}{dt} = -\mu p_0(t) + \lambda p_1(t).$$

Problem 24. Assume that batches of S are ordered when the store becomes empty.

(a)   Show that if $M(t)$ is the expected number of items in the store at time t, then

$$\frac{dM(t)}{dt} = (\lambda + \mu S)p_0(t) - \lambda.$$

(b)   Show that if an equilibrium distribution exists, then $p_n = \mu/(\lambda + \mu S)$, $1 \leqslant n \leqslant S$ and $p_0 = \lambda/(\lambda + \mu S)$.

(c)   Explain briefly why the ratio of empty to non-empty time is approximately $\lambda/(S\mu)$. Show that the expected value of the ratio of empty to non-empty time until the kth re-stocking order is delivered is $k\lambda/[(kS-1)\mu]$. ∎

It may strike the reader as unnecessarily risky to delay placing a bulk order until the store is empty. It should be better to use a buffer stock policy. That is, immediately the stock in hand falls to a certain level, a bulk order is sent out. This is the subject of the next problem.

Problem 25. A store is initially full when it contains S items. Demands are for single items only and arrive in a Poisson stream at an average rate λ per item time. Immediately the store falls to D items, a re-stocking order of $Q = S-D(> D)$ items is made, the delivery time for which is exponential with parameter μ. Even if the store should become empty, no further order is sent out. Show that $p_n(t)$, the probability that the store contains n items at time t, satisfies

$$\frac{dp_n(t)}{dt} = -\lambda p_n(t) + \lambda p_{n+1}(t) + \mu p_{n-Q}(t), \quad Q \leqslant n \leqslant S;$$

$$\frac{dp_n(t)}{dt} = -\lambda p_n(t) + \lambda p_{n+1}(t), \qquad\qquad D < n < Q;$$

$$\frac{dp_n(t)}{dt} = -(\lambda + \mu)p_n(t) + \lambda p_{n+1}(t) \qquad\qquad 0 < n \leqslant D;$$

$$\frac{dp_0(t)}{dt} = -\mu p_0(t) + \lambda p_1(t) \text{ where } p_n(t) \equiv 0, \text{ if } n > S. \quad ∎$$

There is a common-sense reason for insisting, in the above problem, that the bulk re-order $S-D$, should exceed D. This guarantees that whenever it is delivered, the stock level will pass through D and trigger another re-order. A further refinement of the buffer stock policy is effected, if, whenever the store is empty, an additional bulk order for D items is made. However, a small precaution is now necessary. If several supplementary orders of D are made and sufficiently delayed then overflow may take place. This may be 'cured' by cancelling every outstanding D if the order for $S-D$ is delivered first.

6.15 BASIC INVENTORY POLICIES

An inventory policy must state both the occasions for checking the stock level and the re-stocking order consequently to be placed. The existence of a delivery lag prompts a modification in our definition of 'inventory level' which has hitherto been identified as the content of the store. If $c(t)$ is the content of the store and $a(t)$ the amount on order but not yet received at time t, then the inventory level, $I(t)$, shall be defined by

$$I(t) = c(t) + a(t).$$

If backlogging is allowed then $c(t)$ may be negative, indicating that unsatisfied customer demand has been registered. When backlogging is not allowed, $c(t) \geqslant 0$ and we may legitimately describe $c(t)$ as the stock on-hand. There are two basic schedules governing the times at which the inventory level is to be determined.

(1)   In a **periodic review** policy, the inventory level is only determined at regular intervals of time. The amount re-ordered at a review time may be fixed (a *lot*), or, variable. If variable, the order is generally sufficient to restore the inventory level to some desired target maximum (a *replenishment*). This may be achieved by re-ordering $S-I(t)$ at each revew time, where S is the target. If backlogging is not allowed, this is equivalent to re-ordering the amount of stock consumed since the previous review time (see section 6.13). In an important variation, no order is placed at a review time unless $I(t) < s$. We have discussed at some length the optimal choice of s and S when the lead time is zero.

(2)   In a **continuous review** policy, a re-order is made whenever $I(t)$ falls to a specified level, known as the *re-order point*. Such a policy requires the stock to be monitored perpetually and the amount ordered is generally fixed. (See section 6.14). When backlogging is allowed, such a policy is equivalent to re-ordering whenever the content of the store falls by a fixed amount. The so-called **two-bin** policy employs a re-order point. In this, stock is supplied to customers from a bin of capacity Q until it becomes empty, when a replacement order for Q is made to the suppliers. Service then continues from a second bin of capacity $q(< Q)$. The eventual delivery of Q tops up the second bin, the surplus going into the first bin from which service is then resumed. Should the second bin also become exhausted, then one option is to place a supplementary re-stocking order for q. (For a possible hazard see Problem 25.)

There are many variations and combinations of the two basic policies and this variety would make a more general framework and notation advantageous. The reader should consult the specialist literature [5].

REFERENCES

[1] *Distributions in Statistics*, N. Johnston and S. Kotz. Houghton Mifflin, Boston, 1970.

[2] *Studies in the Mathematical Theory of Inventory and Production*, K. Arrow, S. Karlin and H. Scarf. Stanford University Press, 1958.

[3] *Total Positivity*, S. Karlin. Stanford University Press, 1967.

[4] *Applied Probability Models with Optimization Applications*, Sheldon Ross. Holden-Day, San Francisco, 1970.

[5] *Inventory Systems*, E. Naddor. Wiley, New York, 1966.

BRIEF SOLUTIONS AND COMMENTS ON THE PROBLEMS

Problem 1. The economic lot size is $y^* = \sqrt{2.2.9} = 6$ and an order for this is placed every $y^*/m = 6/2 = 3$ units of time. The loss for a period of 3 units is

$$9 + 1.6 + \frac{1}{2}.6.\frac{6}{2} - 5.6 = -6,$$

or -2 per unit of time. Hence $12(-2) = -24$ for twelve units of time. The solution involves the placing of 4 orders.

If we minimise (6.1), we first set $t = y/m$ and obtain $k + (c-r)y + hy^2/2m$. *This* expression is a minimum for $y = m(r-c)/h = 2(5-1)/1 = 8$, and this order is placed every $8/2 = 4$ units of time. The loss over this interval is $(9 + 1.8 + \frac{1}{2}.8.\frac{8}{2} - 5.8) = -7$ and this is certainly less than the loss of -6 for the order of 6 every 3 units of time. However, the loss per unit time is $-7/4 > -6/3$. The fact is that if we are to use (6.1) to find the minimum loss over 12 units of time we must take account not only of the time between re-orders but also of the number of re-orders. Thus, if in time t_0 there are n re-orders, we minimise

$$n(k + cy + hy^2/2m - ry) = t_0\left(\frac{km}{y} + cm + \frac{1}{2}hy - rm\right)$$ where $n = mt_0/y$. But

this means we are to minimise the loss per unit time (6.3).

Problem 2. The minimisation is rather harder than that of Example 2. The reader will soon discover that there is, in general, no common zero for the two partial derivatives. However, using the hint,

$$2km + hS^2 - 2pmS = h\left(S - \frac{pm}{h}\right)^2 - \frac{m^2}{h}\left(p^2 - \frac{2kh}{m}\right)$$

and since $p^2 > 2kh/m$, this expression will be negative if

$$\frac{pm}{h}\left(1 - \sqrt{1-2kh/p^2 m}\right) < S < \frac{pm}{h}\left(1 + \sqrt{1-2kh/p^2 m}\right).$$

Hence for a *fixed* S in this range, the minimum loss per unit time is minimised when $S+s$ is as small as possible, that is when $s = 0$. Now set $s = 0$ and minimise with respect to S. It is easy to show that $\sqrt{2km/h}$ *does* lie in the above interval for S. Other values of S need not be entertained.

Problem 3. Any demand in excess of x is regarded as a cost. Hence the expected cost is

$$cx + r \int_x^\infty (z-x)f(z)\mathrm{d}z.$$

The derivative with respect to x is

$$c - rxf(x) - r \int_x^\infty f(z)\mathrm{d}z + rxf(x)$$

$$= c - r \int_x^\infty f(z)\mathrm{d}z.$$

Hence result.

Problem 4. The only addition to (6.7) is that if the demand is less than x, the remaining $(x-z)$ copies each have salvage value s and the expected recovery is

$s \int_0^x (x-z)f(z)\mathrm{d}z$. The situation is only reasonable if $s < c < r$. Also $(c-s)/(r-s)$

$= [(c-r)/(r-s)] + 1$ which decreases with s.

Problem 5. If $x < y^*$, then $\ell(x) = \ell(y^*,x)$ and the only term in (6.10) which contains x is $c.(y-x)$. Hence $\ell'(x) = -c$. If $x \geqslant y^*$, then $\ell(x) = \ell(x,x)$. In this case, the first term on the right in (6.10) is zero. The derivative of the remaining terms is given by the second two terms in (6.11) with $y = x$. Notice that $\ell'(x)$ is continuous at y^*, since

$$\int_0^{y^*} f(z)\mathrm{d}z = Pr(Z \leqslant y^*) = \frac{p+r-c}{p+r+h}.$$

Problem 6. Following (6.9), the expected loss for ordering up to y is

$$c.(y-x) + \int_y^\infty p(z-y)f(z)\mathrm{d}z + \int_0^y h(y-z)f(z)\mathrm{d}z - r \int_0^y zf(z)\mathrm{d}z - ry$$

$$\int_y^\infty f(z)\mathrm{d}z.$$

Note that $p(.)$ and $h(.)$ are functions. The derivative with respect to y is

$$c - \int_y^\infty p'(z-y)f(z)\,dz + \int_0^y h'(y-z)f(z)\,dz - r \int_y^\infty f(z)\,dz,$$

since common sense dictates $p(0) = h(0) = 0$.

If it is worth ordering anything when the store is empty, the slope is negative at $y = 0$. It is positive for large y since $h'(y-z) > 0$. There is therefore a y^* which makes the first derivative zero. The second derivative with respect to y is

$$\int_y^\infty p''(z-y)f(z)\,dz + \int_0^y h''(y-z)f(z)\,dz + [p'(0) + h'(0) + r]f(y).$$

Each term is positive since $p(.)$, $h(.)$ are convex increasing. Thus y^* gives a minimum cost.

Problem 7

$$h \int_0^y (y-z)f(z)\,dz = h(y-z)F(z)\Big]_0^y + h \int_0^y F(z)\,dz = h \int_0^y F(z)\,dz.$$

$$p \int_y^\infty (z-y)f(z)\,dz = -\Big[p(z-y)\{1-F(z)\}\Big]_y^\infty + p \int_y^\infty [1-F(z)]\,dz$$

$$= pE(Z) - p \int_0^y [1-F(z)]\,dz.$$

$$-r \int_0^y zf(z)\,dz - ry \int_y^\infty f(z)\,dz = -ry + r \int_0^y F(z)\,dz. \text{ Hence result.}$$

Use

$$\int_0^y F(z)\,dz < \int_0^y F(y^*)\,dz = yF(y^*), \text{ for } y < y^*.$$

We conclude that if $c < p + r$, $\ell(y,0) < pE(Z)$ for $y < y^*$. Note that $pE(Z)$ is the expected loss if nothing is ordered.

Problem 8

$$L(y) = 1. \int_0^y (y-z)\,dz + 1. \int_y^1 (z-y)\,dz$$

$$= y^2 - y + \frac{1}{2}.$$

Now $E(Z) = \frac{1}{2}$, hence

$$\ell(y,x) = \frac{1}{64} + \frac{1}{2}(y-x) + y^2 - y + \frac{1}{2} - \frac{3}{8}, \quad y > x.$$

$$\ell(x,x) = x^2 - x + \frac{1}{2} - \frac{3}{8}.$$

$\frac{\partial}{\partial y}\left[\ell(y,x)\right] = \frac{1}{2} + 2y - 1$ and there is a minimum at $y = \frac{1}{4}$ which is attainable

if $x < \frac{1}{4}$. Hence $y^* = \frac{1}{4}$ and $\ell\left(\frac{1}{4},x\right) = \frac{5}{64} - \frac{x}{2}$. The remaining question is to

locate those x such that $\ell(x,x) \leqslant \ell\left(\frac{1}{4},x\right)$. That is, $x^2 - x + \frac{1}{8} \leqslant (5 - 32x)/64$

or $\left(x - \frac{1}{8}\right)\left(x - \frac{3}{8}\right) \leqslant 0$. Hence $x^* = \frac{1}{8}$, and the optimal policy is:

if $x < \frac{1}{8}$, bring the inventory up to $1/4$,

if $x \geqslant \frac{1}{8}$ order nothing,

The minimum expected loss, $\ell(x)$, is

$$\ell(x) = \frac{5}{64} - \frac{x}{2}, \qquad x < \frac{1}{8},$$

$$\ell(x) = x^2 - x + \frac{1}{8}, \qquad x \geqslant \frac{1}{8}.$$

Problem 9

$E(Z) = 8, p + r = 2$, hence if $0 \leqslant x < 4$,

$$\ell_2(x) = 1.(4-x) + 2\int_0^4 \left(\frac{4-z}{16}\right)dz + 2\int_4^{16}\left(\frac{z-4}{16}\right)dz - \frac{3}{2}E(Z),$$

$$= (4-x) + 1 + 9 - 12 = 2-x. \text{ While, if } x \geqslant 4,$$

$$\ell_2(x) = 2\int_0^x\left(\frac{x-z}{16}\right)dz + 2\int_x^{16}\left(\frac{z-x}{16}\right)dz - \frac{3}{2}E(Z),$$

$$= \frac{x^2}{16} + \frac{(16-x)^2}{16} - 12 = \frac{x^2}{8} - 2x + 4.$$

Note that $\ell_2(x)$ is continuous at $x = 4$, when it assumes the value -2.

Problem 10. If initially $x_1 > 5$, nothing is ordered at the start of the first period.

(i) $z_1 > x_1$, order 4. Total contributions to both periods

$$2\int_{x_1}^{16}\left(\frac{z_1-x_1}{16}\right)dz_1 + \alpha\int_{x_1}^{16}\frac{2}{16}dz_1.$$

(ii) $x_1 - 4 \leqslant z_1 \leqslant x_1$, level at end of first period is $x_1 - z_1$, order $4 - (x_1 - z_1)$
$= z_1 - (x_1 - 4)$. Total contribution to both periods

$$2 \int_{x_1-4}^{x_1} \left(\frac{x_1-z_1}{16}\right) dz_1 + \alpha \int_{x_1-4}^{x_1} \left(\frac{2-x_1+z_1}{16}\right) dz_1.$$

(iii) $0 \leqslant z_1 < x_1 - 4$, order nothing at the start of the second period. Total contribution to both periods

$$2 \int_0^{x_1-4} \left(\frac{x_1-z_1}{16}\right) dz_1 + \alpha \int_0^{x_1-4} \left\{\frac{(x_1-z_1)^2 - 16(x_1-z_1)+32}{8.16}\right\} dz_1.$$

From the total of (i), (ii), (iii) subtract 12 which compensates for the incorporation of lost revenue in the penalty for the first period.

Problem 11. The minimum expected loss for the whole n periods consists of the minimum of the sum of the expected loss for the first period and the discounted loss for the remaining $n-1$ periods. $\ell_{1,n}(x)$ is

$$\min_{y \geqslant x} \left[c.(y-x)+L(y)-rE(Z)+\alpha\left\{\ell_{2,n-1}(0) \int_y^\infty f(z)dz + \right.\right.$$

$$\left.\left. \int_0^y \ell_{2,n-1}(y-z)f(z)dz\right\}\right].$$

Now take the limit as $n \to \infty$. If $G'(y^*) = 0$, then

$$\ell(x) = -cx + G(y^*) - rE(Z) \Rightarrow \ell'(x) = -c, \quad (x < y^*)$$

and

$$\ell(x) = -cx + G(x) - rE(Z) \Rightarrow \ell'(x) = -c+G'(x), \quad (x \geqslant y^*).$$

But $G'(y^*) = 0$, implies $c + L'(y^*) + \alpha \int_0^{y^*} \ell'(y^*-z)f(z)dz = 0$. Further, $0 < z < y^* \Rightarrow y^*-z < y^*$, hence

$$c + L'(y^*) - \alpha c \int_0^{y^*} f(z)dz = 0.$$

Problem 12. If $y_1 < 1/8$, then $y_1 - z_1 < 1/8$. Hence $\ell_2'(y_1 - z_1) = -1/2$ for $y_1 < 1/8$. It follows that

$$G'(y_1) = \frac{1}{2} + 2y_1 - 1 + \alpha \int_0^{y_1} (-1/2)dz_1 = 2y_1 - \frac{1}{2} - \frac{\alpha y_1}{2}.$$

Since $0 < \alpha < 1$, $G'(y_1)$ is negative for $0 < y_1 < 1/8$.

Problem 13

$$G'(y_1) = c + L'(y_1) + \alpha \int_0^{y_1} \ell_2'(y_1 - z_1)f(z_1)dz_1,$$

$$= c + L'(y_1) + \alpha \left[\int_{y_1-x_2^*}^{y_1} -c.f(z_1)dz_1 + \right.$$

$$\left. \int_0^{y_1-x_2^*} L'(y_1-z_1)f(z_1)dz_1 \right]$$

$$= c + L'(y_1) - \alpha \{ c[F(y_1)-F(y_1-x_2^*)] - $$

$$\int_0^{y_1-x_2^*} L'(y_1-z_1)f(z_1)dz_1 \}.$$

After differentiating again, and re-arranging we can write $G_1''(y_1)$ as

$$L''(y_1) + \alpha \left[f(y_1-x_2^*) \{ c+L'(x_2^*) \} -cf(y_1) + \int_0^{y_1-x_2^*} L''(y_1-z_1)f(z_1)dz_1 \right].$$

The term $c + L'(x_2^*)$ is negative, and even if $L(.)$ is convex so that $L''(y_1)>0$, there is no guarantee that $G''(y_1)$ is positive.

Problem 14

$$G(y_1) = y_1 + L_1(y_1) - 15 + 1. \int_0^{15} \frac{\ell_2(y_1-z_1)}{15} dz_1.$$

If $y_1-z_1 > 6$, then $\ell_2(y_1-z_1)=L_2(y_1-z_1)-15$, while if $y_1-z_1 < 6$, $\ell_2(y_1-z_1)$ $= 6-(y_1-z_1)+L_2(6) - 15$.

Hence, substituting from (6.45) and (6.46),

$$\int_0^{15} \frac{\ell_2(y_1-z_1)}{15} dz_1 = \int_0^{y_1-6} \left[\frac{(y_1-z_1)^2}{6} -3(y_1-z_1) + \frac{15}{2} \right] \frac{1}{15} dz_1 + $$

$$\int_{y_1-6}^{15} \left[\frac{3}{2} - (y_1-z_1) \right] \frac{1}{15} dz_1.$$

The derivative of the expression on the right-hand side is

$$\int_0^{y_1-6} \left[\frac{y_1-z_1}{45} - \frac{3}{15} \right] dz_1 - \frac{9}{30} + \int_{y_1-6}^{15} \frac{-1}{15} dz_1 + \frac{9}{30},$$

which at $y_1 = 6$, has the value -1. Hence, since $\alpha = 1$, $G'(6) = 1 + L_1'(6) - 1 = 1/5 > 0$. It is readily shown that $G''(y_1) > 0$ for $6 < y_1 < 15$ and the result follows.

Problem 15. We have $\ell_1(x_1) = L_1(x_1)-15 + 1. \int_0^{15} \frac{\ell_2(x_1-z_1)}{15} dz_1$. From the solution to Problem 14,

$$\int_0^{x_1-6} \frac{\ell_2(x_1-z_1)}{15} dz_1 = \int_0^{x_1-6} \left[\frac{(x_1-z_1)^2}{6} - 3(x_1-z_1) + \frac{15}{2} \right] \frac{1}{15} dz_1$$

$$= \frac{1}{15} \left[\frac{x_1^3}{18} - \frac{3x_1^2}{2} + \frac{15x}{2} - 3 \right].$$

$$\int_{x_1-6}^{15} \frac{\ell_2(x_1-z_1)}{15} dz_1 = \int_{x_1-6}^{15} \left[\frac{3}{2} - (x_1-z_1) \right] \frac{1}{15} dz_1$$

$$= \frac{1}{15} \left[\frac{x_1^2}{2} - \frac{33x_1}{2} + 126 \right].$$

For $\alpha = 1$, $L_1(x_1) = \frac{x_1^2}{15} + \frac{(15-x_1)^2}{30}$.

Problem 16. Here the situation in the first period is slightly different. If $y_1 > 15$, then there is a holding penalty but there is no backlogged revenue since the demand is less than 15.

$$\ell_1(x_1) = \min_{y_1 \geqslant x_1} \left[(y_1-x_1) + 2 \int_0^{15} \frac{(y_1-z_1)}{15} dz_1 - 2 \int_0^{15} \frac{z_1}{15} dz_1 + \right.$$

$$\left. \alpha \int_0^{15} \ell_2(y_1-z_1) dz_1 \right]$$

$$= -x_1 - 30 + \min_{y_1 \geqslant x_1} [G(y_1)].$$

After collecting terms from the solution to Problem 15, since $y_1 - 6 < 15$,

$$G(y_1) = 3y_1 + \frac{\alpha}{15} \left[\frac{y_1^3}{18} - y_1^2 - 9y_1 + 123 \right].$$

Hence $G'(y_1), G''(y_1)$ are both positive for $15 < y_1 < 21$.

Problem 17. The expected loss for the first period is $cy^* + L_1(y^*) - rE(Z)$. At the end of each period, the contents are y^*-Z, hence Z must be ordered. The loss for each subsequent period is $cE(Z) + L_1(y^*) - rE(Z)$. Hence the total (discounted) expected loss is

$$cy^* + L_1(y^*) - rE(Z) + \sum_2^\infty \alpha^{i-1} [cE(Z) + L_1(y^*) - rE(Z)].$$

Differentiate with respect to y^* to obtain result.

Problem 18. If $x_1 < 3$, then $x_2 = x_1-z_1 < 3$ for all z_1. Hence from (6.50) $\ell_2(x_1-z_1) = 3+z_1-x_1$. Thus (6.53) becomes

$$\frac{x_1{}^2}{15} + \frac{(15-x_1)^2}{30} - 15 + \int_0^{15} \frac{(3+z_1-x_1)}{15} \, dz_1 = \frac{x_1{}^2}{10} - 2x_1 + 3.$$

For the evaluation we give some partial results.

$$L_1(11/2) = \frac{121}{60} + \frac{361}{120} = \frac{201}{40}.$$

$$\int_0^{15} \ell_2(5\cdot5-z_1) \, dz_1 = \int_{2\cdot5}^{15} (3+z_1-5\cdot5) \, dz_1 + \int_0^{2\cdot5} \left[\frac{(5\cdot5-z_1)^2}{6} \right.$$

$$\left. -3(5\cdot5-z_1) + \frac{15}{2} \right] dz_1$$

$$= \frac{625}{8} - \frac{775}{144}.$$

Problem 19

$\pi_j = \displaystyle\sum_0^2 \pi_i p_{ij}$ since $p_i^{(n-1)} \to \pi_i$ and $p_j^{(n)} \to \pi_j$. By substitution, if $j = 0$, the right-hand side is proportional to

$$(1-\phi_0)^2 (1-\phi_0-\phi_1) + \phi_1(1-\phi_0) + \phi_0(1-\phi_0)(1-\phi_0-\phi_1)$$
$$= (1-\phi_0) [(1-\phi_0-\phi_1)\{(1-\phi_0 + \phi_0)\} + \phi_1]$$
$$= (1-\phi_0) [(1-\phi_0-\phi_1) + \phi_1] = (1-\phi_0)^2. \text{ Similarly for } j = 1,2.$$

Alternatively,

$$\pi_0 = \pi_0(1-\phi_0-\phi_1) + \pi_1(1-\phi_0) + \pi_2(1-\phi_0-\phi_1) \quad (1)$$
$$\pi_1 = \pi_0\phi_1 + \pi_1\phi_0 + \pi_2\phi_1 \quad (2)$$
$$\pi_2 = \pi_0\phi_0 + \pi_2\phi_0 \quad (3)$$

From equation (3), $\pi_2 = \pi_0\phi_0/(1-\phi_0)$; substitute in equation (2).

$$\pi_1(1-\phi_0) = \pi_0\phi_1 + \frac{\pi_0\phi_1\phi_0}{1-\phi_0} = \frac{\pi_0\phi_1}{1-\phi_0}$$

$$\Rightarrow \pi_1 = \frac{\pi_0\phi_1}{(1-\phi_0)^2}. \text{ Now use } \pi_0 + \pi_1 + \pi_2 = 1.$$

Problem 20. There will be 2 items at the beginning of the nth period if there were:

2 at the beginning of the $(n-1)$th period, a demand for at *least* two items and an order was placed;

2 at the beginning of the $(n-1)$th period and no demand;
1 at the beginning of the $(n-1)$th period, a demand for at *least* one item and an order was placed.

Thus, if $\rho_2^{(n)}$ is the probability that there are 2 items at the start of the nth period,

$$\rho_2^{(n)} = \rho_2^{(n-1)}(1-\phi_0-\phi_1) + \rho_1^{(n-1)}(1-\phi_0) + \rho_2^{(n-1)}\phi_0.$$

Hence, in the limit, $\rho_2\phi_1 = \rho_1(1-\phi_0)$. Now use $\rho_1 + \rho_2 = 1$.

Problem 21. In (6.58), set $j = 1$.

$$p_1^{(n)} = p_0^{(n-1)}p_{01} + p_1^{(n-1)}p_{11} + p_2^{(n-1)}p_{21}.$$

Substitute $p_2^{(n-1)} = 1 - p_0^{(n-1)} - p_1^{(n-1)}$ and we have the required result since the coefficient of $p_0^{(n-1)}$ after collecting terms is zero. After repeated application we arrive at

$$p_1^{(n)} = \frac{\phi_1[1-(\phi_0-\phi_1)^n]}{1-\phi_0+\phi_1}$$

$$\lim_{n \to \infty}[p_1^{(n)}] = \frac{\phi_1}{1-\phi_0+\phi_1} = \pi_1.$$

Problem 22. Note that S sales are made if there are S or more demands.

Problem 23. Put $r = S$ in equation (6.59)

$$p_S^{(n)} = \phi_0 \sum_{i=0}^{S-1} p_{S-i}^{(n-1)} + p_0^{(n-1)},$$

$$= \phi_0[1-p_0^{(n-1)}] + p_0^{(n-1)}, \text{ and } \phi_0 = \exp(-\lambda t_0).$$

Problem 24

(a) $$\frac{dM(t)}{dt} = \frac{d}{dt}\left[\sum_{n=1}^{S} np_n(t)\right] = \sum_{1}^{S} n\frac{dp_n(t)}{dt}$$

$$= \sum_{1}^{S-1} n\frac{dp_n(t)}{dt} + S\frac{dp_S(t)}{dt}$$

$$= \lambda \sum_{1}^{S-1} [np_{n+1}(t)-np_n(t)] - \lambda Sp_S(t) + \mu Sp_0(t)$$

$$= \lambda \sum_{1}^{S-1} [(n+1)p_{n+1}(t)-p_{n+1}(t) - np_n(t)] - \lambda Sp_S(t)+\mu Sp_0(t)$$

$$= \lambda\{Sp_S(t) - p_1(t)\} - \lambda\{1 - p_0(t) - p_1(t)\} - \lambda Sp_S(t) + \mu Sp_0(t)$$

$$= (\lambda + \mu S)p_0(t) - \lambda.$$

(b) If an equilibrium distribution exists, set $\dfrac{dp_n(t)}{dt} = 0$ and $p_n(t) = p_n$ in the equations

$$0 = -\lambda p_n + \lambda p_{n+1}, \qquad\qquad 1 \leqslant n \leqslant S-1$$

$$0 = -\lambda p_S + \mu p_0,$$

$$0 = -\mu p_0 + \lambda p_1.$$

Hence $p_n = p_{n+1}$ $(1 \leqslant n \leqslant S-1)$, $p_S = (\mu/\lambda)p_0$, $p_1 = (\mu/\lambda)p_0$. Using $\sum\limits_{0}^{S} p_n = 1$, we obtain $p_n = \mu/(\lambda + \mu S)$, $p_0 = \lambda/(\lambda + \mu S)$.

(c) The expected time between demands is $1/\lambda$, hence the expected time to empty a full store is S/λ. Expected time for batch order to arrive is $1/\mu$. Hence result. The time to empty a full store has a $\Gamma(S, \lambda)$ distribution. The total non-empty time has a $\Gamma(kS, \lambda)$ distribution. The total empty time has a $\Gamma(k, \mu)$ distribution.

E(empty time/non-empty time) $= E$(empty time)E(1/non-empty time).

The mean of the reciprocal is *not* the reciprocal of the mean. If T is $\Gamma(kS, \lambda)$ then $E(1/T) = \lambda/(kS-1)$.

Problem 25. Consider $p_n(t+\delta t)$.

(a) If $Q \leqslant n < S$, then at t either the contents were (i) n and no demand, (ii) $n+1$ and one demand, (iii) $n-Q$ and a delivery at $t+\delta t$.

$$p_n(t+\delta t) = p_n(t)(1-\lambda\delta t) + p_{n+1}(t)\lambda\delta t + p_{n-Q}(t)\mu\delta t.$$

Similarly $p_S(t+\delta t) = p_S(t)(1-\lambda\delta t) + p_D(t)\mu\delta t$, which agrees with $n = S$ provided $p_{S+1}(t) \equiv 0$.

(b) If $D < n < Q$, then $n-Q < 0$ so that the contents at t cannot have been $n-Q$.

$$p_n(t+\delta t) = p_n(t)(1-\lambda\delta t) + p_{n+1}(t)\lambda\delta t.$$

(c) If $0 < n < D$ there must have been an order out for Q.

$$p_n(t+\delta t) = p_n(t)(1-\lambda\delta t)(1-\mu\delta t) + p_{n+1}(t)\lambda\delta t(1-\mu\delta t).$$

$$p_D(t+\delta t) = p_D(t)(1-\lambda\delta t)(1-\mu\delta t) + p_{D+1}(t)\lambda\delta t.$$

(d) $p_0(t+\delta t) = p_0(t)(1-\mu\delta t) + p_1(t)\lambda\delta t(1-\mu\delta t).$

Bear in mind that if the store is empty, no demand can be satisfied.

Probability Generating Functions

The following result should be engraved on the hearts of our readers. 'If a coin is tossed independently n times and the probability of a head on each toss is p, then the probability of the occurrence of just r heads is $\binom{n}{r} p^r(1-p)^{n-r}$'. The derivation depends only on noting the fact that any particular sequence of r heads and $n-r$ tails has probability $p^r(1-p)^{n-r}$, and then observing that the number of distinct ways in which such a sequence can be arranged is $\binom{n}{r}$. This counting of the number of ways in which an event can occur is seldom so simple as selecting r places from n and so we seek a more general process.

We first observe that $\binom{n}{r}$ is also the coefficient of θ^r in the binomial expansion of $(1 + \theta)^n$. We may say that $(1 + \theta)^n$ is a **generating function** for the sequence $\binom{n}{0}, \ldots \binom{n}{r}, \ldots \binom{n}{n}$. Many properties of the binomial coefficients may easily be established by operating on the function $(1 + \theta)^n$.

Example 1

If $(1 + \theta)^n = \sum_{0}^{n} \binom{n}{r} \theta^r$, then

(a) by setting $\theta = 1$, $\sum_{0}^{n} \binom{n}{r} = 2^n$;

(b) differentiating with respect to θ and then setting $\theta = 1$,

$$\sum_{1}^{n} r\binom{n}{r} = n \, 2^{n-1};$$

(c) since $(1 + \theta)^n = (1 + \theta)^m (1 + \theta)^{n-m}$, then for $0 < m < n$,

$$\sum_{k=0}^{\min(r,m)} \binom{m}{k}\binom{n-m}{r-k} = \binom{n}{r}.$$

The way in which such a generating function is constructed may be seen more clearly from the following consideration. Consider the placing of r *similar* balls into n *different* boxes so that no box receives more than one ball ($r \leqslant n$). Thus for each box we 'indicate' that it either receives a ball or it does not. The record of the placements will consist of a sequence of ones and zeros. Each such sequence will correspond to one term in the expansion

$$(1+\theta)^n = \prod_1^n (1+\theta) = (1+\theta)(1+\theta)\ldots(1+\theta).$$

Each term in the product is found by selecting 1 or θ from each of the n factors $(1+\theta)$ and multiplying them together. Now let the ith box correspond to the ith factor. If the box is empty we select $1 = \theta^0$ and if it contains a ball we select $\theta = \theta^1$ from the corresponding factor. This suggests how we may solve the more difficult problem of the number of different assignments when there is no restriction on the number of balls in a box. We include a factor $(\theta^0+\theta^1+ \ldots +\theta^r)$ for each of the n boxes. Then we need the coefficient of θ^r in the expansion of $(1+\theta+\theta^2 + \ldots +\theta^r)^n$. Now,

$$(1+\theta+\theta^2+ \ldots +\theta^r)^n = \left(\frac{1-\theta^{r+1}}{1-\theta}\right)^n$$

$$= (1-\theta^{r+1})^n(1-\theta)^{-n}$$

$$= (1-\theta^{r+1})^n(1+n\theta+ \ldots + \frac{n(n+1)\ldots(n+r-1)\theta^r}{r!} \ldots),$$

and the coefficient of θ^r is $\binom{n+r-1}{r}$. We are in no difficulty about the expansion of $(1-\theta)^{-n}$, since θ is a dummy variable and we can accept restrictions on its magnitude.

Problem 1. By considering the expression $(\theta+\theta^2+ \ldots +\theta^r)^n$, show that, when $r \geqslant n$, the number of distinguishable assignments of r similar balls into n different boxes so that each box contains at least one ball is $\binom{r-1}{r-n}$. ■

Only a small extension of the ideas just discussed is required to find the distribution of the sum of independent random variables which have positive probability on the values $0,1,2,\ldots$. For suppose $Pr(X_1 = i) = p_{1i}(i = 0,1,2,\ldots), Pr(X_2 = j) = p_{2j}(j = 0,1,2,\ldots)$, then

$$Pr(X_1+X_2 = k) = \sum_{i=0}^{k} Pr(X_1 = i \text{ and } X_2 = k-i)$$

$$= \sum_{0}^{k} Pr(X_1 = i)Pr(X_2 = k-i)$$

$$= \sum_{0}^{k} p_{1,i} \, p_{2,k-i}.$$

But this last expression is the coefficient of θ^k in the product

$$(p_{10}+p_{11}\theta+p_{12}\theta^2 \ldots)(p_{20}+p_{21}\theta+p_{22}\theta^2 \ldots).$$

Since the coefficients are probabilities which sum to 1, the series in each bracket converges for $|\theta| \leqslant 1$, and these are called the **probability generating functions** for X_1, X_2, written $G_{X_1}(\theta), G_{X_2}(\theta)$ respectively. Our result states that

$$G_{X_1}(\theta)G_{X_2}(\theta) = G_{X_1+X_2}(\theta).$$

From inspection, this identity corresponds to the obvious equality

$$E(\theta^{X_1})E(\theta^{X_2}) = E(\theta^{X_1+X_2}),$$

when X_1, X_2 are independent.

Example 2
X has the Poisson distribution with parameter λ.

$$G_X(\theta) = E(\theta^X) = \sum_{0}^{\infty} \theta^x \lambda^x e^{-\lambda}/x!$$

$$= e^{-\lambda}e^{\lambda\theta} = e^{-\lambda+\lambda\theta}.$$

Thus if X_1 has the Poisson distribution with parameter λ_1, and the independent random variable X_2 has another Poisson distribution with paramete λ_2, then

$$G_{X_1+X_2}(\theta) = G_{X_1}(\theta)G_{X_2}(\theta) = \exp(-\lambda_1+\lambda_1\theta)\exp(-\lambda_2+\lambda_2\theta)$$

$$= \exp[-(\lambda_1+\lambda_2) + (\lambda_1+\lambda_2)\theta].$$

But this is the p.g.f. of another Poisson distribution with parameter $(\lambda_1+\lambda_2)$.

Problem 2. The distribution of X, is Poisson with parameter λT. If T has an exponential distribution with parameter μ, by considering

$$\int_0^\infty E[\theta^X|t] \, \mu \exp(-\mu t)dt,$$

find the marginal distribution of X. ∎

Example 3
X has the binomial distribution with parameters n,p.

$$G_X(\theta) = E(\theta^X) = \sum_0^n \binom{n}{x}(p\theta)^x(1-p)^{n-x}$$

$$= [p\theta+(1-p)]^n.$$

Hence if X_1 has the binomial distribution with parameters n_1,p and X_2 has an independent binomial distribution with parameters n_2,p then

$$G_{X_1+X_2}(\theta) = G_{X_1}(\theta)G_{X_2}(\theta) = [p\theta+(1-p)]^{n_1}[p\theta+(1-p)]^{n_2}$$

$$= [p\theta+(1-p)]^{n_1+n_2},$$

which corresponds to a binomial distribution with parameters n_1+n_2,p.

Problem 3. The distribution of X, given n, is binomial with parameters n,p_1. The distribution of N is binomial with parameters m,p_2. Show that the marginal distribution of X is binomial with parameters m,p_1p_2. ∎

Another useful feature of a probability generating function is that it can be used to find the moments of a distribution, when these exist. For if $G_X(\theta) = \sum_0^\infty p_n\theta^n$, then $G_X'(\theta) = \sum_1^\infty np_n\theta^{n-1}$. If $\lim_{\theta\to1}\left(\sum_1^\infty np_n\theta^{n-1}\right)$ exists, then it is $E(X)$ and is also equal to $\lim_{\theta\to1}[G_X'(\theta)]$. Similarly $G_X''(\theta) = \sum_2^\infty n(n-1)p_n\theta^{n-2}$ and if

$$\lim_{\theta\to1}\left[\sum_2^\infty n(n-1)p_n\theta^{n-2}\right]$$

exists, it is $E[X(X-1)]$ and is also $\lim_{\theta\to1}[G_X''(\theta)]$.

Problem 4. Use the probability generating function to obtain the mean and variance of (a) the Poisson distribution (b) the binomial distribution. ∎

BRIEF SOLUTIONS AND COMMENTS ON THE PROBLEMS

Problem 1

$$(\theta+\theta^2+\ldots+\theta^r)^n = \left(\frac{\theta-\theta^{r+1}}{1-\theta}\right)^n = \theta^n(1-\theta^r)^n(1-\theta)^{-n}.$$

We require the coefficient of θ^{r-n} in the expansion of $(1-\theta)^{-n}$. This is easily found to be $\binom{r-1}{n-1}$. Alternatively put one ball in each box, leaving $r-n$. There are $\binom{n+r-n-1}{r-n} = \binom{r-1}{r-n} = \binom{r-1}{n-1}$ different assignments of these $r-n$ balls with no restriction.

Problem 2. The p.g.f. of a Poisson distribution with parameter λt is $\exp(-\lambda t + \lambda t\theta)$. Hence the p.g.f. of the marginal distribution of X is

$$\int_0^\infty \exp(-\lambda t + \lambda t\theta)\mu \exp(-\mu t)\mathrm{d}t = \mu/(\lambda+\mu-\lambda\theta)$$

$$= \frac{\mu}{\lambda+\mu}\left(1 - \frac{\lambda}{\lambda+\mu}\theta\right)^{-1}.$$

$Pr[X = n]$ is the coefficient of θ^n, which is $\mu\lambda^n/(\lambda+\mu)^{n+1}$.

Problem 3. The p.g.f. of the distribution of X given n is $(p_1\theta+1-p_1)^n$. Hence the p.g.f. of the unconditional distribution of X is

$$\sum_{n=0}^m (p_1\theta+1-p_1)^n\binom{m}{n}p_2^n(1-p_2)^{m-n} = [(p_1\theta+1-p_1)p_2+1-p_2]^m$$

$$= (p_1p_2\theta+1-p_1p_2)^m. \text{ This is the p.g.f. of a binomial distribution with parameters } m, p_1p_2.$$

Problem 4

(i) $G(\theta) = \exp(-\lambda+\lambda\theta)$, hence $G'(1) = \lambda$, $G''(1) = \lambda^2$.
$V(X) = E[X(X-1)] + E(X) - E^2(X) = \lambda^2+\lambda-\lambda^2 = \lambda$.
(ii) $G(\theta) = (p\theta+1-p)^n$. $G'(1) = np$.
$G''(1) = n(n-1)p^2$. $V(X) = n(n-1)p^2+np-(np)^2 = np(1-p)$.

The Laplace Transform

The reader will recall that a distribution will only have a moment generating function if moments of all orders exist. However, most of the random variables discussed in this text are non-negative and in these cases another choice of weighting function will give a transform which *will* exist. More generally, let $g(t)$ be a continuous function specified for $t \geqslant 0$. Then the Laplace transform, $g^*(s)$, is defined to be

$$g^*(s) = \int_0^\infty e^{-st} g(t) \, dt, \qquad (A2.1)$$

providing the integral exists. The transform, also written $L[g(t);s]$, is of course a function of s and does not usually exist for all values of s. For our purpose, there will be no serious loss of generality if we restrict attention to real values of s.

In particular, if $f(t)$ is the p.d.f. of a non-negative continuous random variable, T, then

$$f^*(s) = \int_0^\infty e^{-st} f(t) \, dt. \qquad (A2.2)$$

In this case,

$$f^*(0) = \int_0^\infty f(t) \, dt = 1, \qquad (A.2.3)$$

and since, for $s > 0$ we have $e^{-st} \leqslant 1$ and since $f(t) \geqslant 0$, it follows that $f^*(s)$ exists for $s \geqslant 0$. We also note that, by equation (A2.2),

$$f^*(s) = E[\exp(-sT)]. \qquad (A2.4)$$

Example 1 $g(t) = ae^{bt}, t \geqslant 0.$

$$g^*(s) = \int_0^\infty e^{-st} ae^{bt} \, dt$$

$$= a \int_0^\infty e^{t(b-s)} \, dt$$

$$= a \lim_{x \to \infty} \int_0^x e^{t(b-s)} \, dt$$

$$= a \lim_{x \to \infty} \left[\frac{e^{t(b-s)}}{b-s} \right]_0^x , s \neq b$$

$$= a \lim_{x \to \infty} \frac{e^{x(b-s)}}{b-s} + \frac{a}{s-b} .$$

Clearly the limit exists and is zero provided that $s > b$ and $g^*(s)$ is then

$$\frac{a}{s-b} .$$
(A2.5)

The limit does not exist for $s \leq b$.

In particular, if T has the exponential distribution with p.d.f. $f(t) = \lambda \exp(-\lambda t)$, then from (A2.5)

$$f^*(s) = \frac{\lambda}{\lambda+s} ,$$
(A2.6)

and exists for $s > -\lambda$ but as $\lambda > 0$ this infinite interval includes the origin.

Example 2 $g(t) = at^\alpha, t \geqslant 0, \alpha > -1$

$g^*(s) = a \int_0^\infty e^{-st} t^\alpha \, dt$, changing the variable to $x = st$,

$$= \frac{a}{s^{\alpha+1}} \int_0^\infty x^\alpha e^{-x} \, dx, \quad s > 0$$

$$= \frac{\Gamma(\alpha+1)}{s^{\alpha+1}}, \text{ provided } \alpha + 1 > 0.$$

When α is a positive integer, m,

$$g^*(s) = \frac{a(m!)}{s^{m+1}} .$$
(A2.7)

Also, when $\alpha = 0$,

$$g^*(s) = \frac{a}{s} .$$
(A2.8)

Example 3 $g(t) = a \cos(bt), t \geqslant 0$

$$g^*(s) = a \int_0^\infty e^{-st} \cos(bt) dt.$$

$g*(s)$ exists if $s > 0$ and, after integrating by parts,

$$g*(s) = \frac{a}{s} - \frac{ab}{s} \int_0^\infty e^{-st} \sin(bt)dt$$

$$= \frac{a}{s} - \frac{ab^2}{s^2} \int_0^\infty e^{-st} \cos(bt)dt$$

$$= \frac{a}{s} - \frac{b^2}{s^2} g*(s)$$

whence

$$g*(s) = as/[s^2 + b^2].$$

Problem 1. Find the Laplace transforms of (i) $a \sin(bt)$, (ii) $t^n e^{-bt}/n!$, (iii) $e^{-at} \cos(bt)$. ■

There are a number of properties which allow the Laplace transform to be identified when certain modifications are made to the original function, e.g. change of location, change of scale.

Example 4
If $h(t) = e^{at}g(t)$, $t \geq 0$, then $h*(s) = g*(s-a)$. For

$$h*(s) = \int_0^\infty e^{-st} e^{at}g(t)dt$$

$$= \int_0^\infty e^{-t(s-a)} g(t)dt$$

$$= g*(s-a).$$

If $g*(s)$ exists for $s \geq s_0$, then $h*(s)$ exists for $s \geq s_0 + a$.

Problem 2. If $h(t) = g(t-a)$, $t \geq a$, $h(t) = 0$, $t < a$, show that $h*(s) = \exp(-as) g*(s)$. ■

Problem 3. If $h(t) = g(at)$ show that $h*(s) = \frac{1}{a} g* \left(\frac{s}{a}\right)$. ■

Example 5 $h(t) = g'(t)$

$$h*(s) = \int_0^\infty e^{-st} g'(t)dt,$$

and integrating by parts this

$$= \left[e^{-st}g(t)\right]_0^\infty + s \int_0^\infty e^{-st}g(t)dt.$$

Now, if $\lim_{t \to \infty} [e^{-st}g(t)] = 0$, i.e. $g(t)$ does not increase too rapidly, and $g^*(s)$ exists, $h^*(s) = -g(0) + sg^*(s)$.

Problem 4. If $h(t) = \int_0^t g(x)dx$, show that $h^*(s) = g^*(s)/s$. ■

Problem 5. If $h_n(t) = t^n g(t)$, show that $\dfrac{dh_n^*(s)}{ds} = -h_{n+1}^*(s)$. Show further that

$$h_{-1}^*(s) = h_{-1}(0) - \int_0^s g^*(x)dx. \quad ■$$

STATISTICAL APPLICATIONS

We have already observed that if T is a non-negative random variable, then the Laplace transform of the probability density function always exists at least for $s \geqslant 0$.

Example 6

If T has the gamma distribution with parameters α, λ then

$$f(t) = \frac{\lambda(\lambda t)^{\alpha-1} \exp(-\lambda t)}{\Gamma(\alpha)}, t \geqslant 0.$$

Then

$$f^*(s) = \int_0^\infty \frac{\lambda(\lambda t)^{\alpha-1} \exp[-t(\lambda+s)]}{\Gamma(\alpha)} dt.$$

After changing the variable to $y = t(\lambda+s)$,

$$f^*(s) = \frac{\lambda^\alpha}{(\lambda+s)^\alpha} \int_0^\infty \frac{y^{\alpha-1} e^{-y}}{\Gamma(\alpha)} dy$$

$$= \left(\frac{\lambda}{\lambda+s}\right)^\alpha. \tag{A2.9}$$

If $\alpha = 1$, then equation (A2.9) is equation (A2.6), as it should be.

Problem 6. If $f(t)$ is the p.d.f. of a non-negative random continuous variable and the corresponding distribution function is $F(t)$, show that

$$F^*(s) = \frac{F(0)}{s} + \frac{f^*(s)}{s}.$$

For a continuous random variable, $F(0) = 0$ and the equation reduces to $F^*(s) = f^*(s)/s$. ■

We exploit the identification of the Laplace transform as an expectation to obtain some further results. If T_1, T_2 are non-negative and *independent* random variables with p.d.f.s $f_1(.), f_2(.)$, then

$$E\left[e^{-(T_1 + T_2)s}\right] = E(e^{-T_1 s}) E(e^{-T_2 s})$$

$$= f_1^*(s) f_2^*(s). \tag{A2.10}$$

That is, the Laplace transform of the p.d.f. of the distribution of $T_1 + T_2$ is the product of the Laplace transforms of the p.d.f.s of the distributions of T_1, T_2.

Example 7

If T_1 has an exponential distribution with parameter λ_1 and the independent random variable T_2 has an exponential distribution with parameter λ_2, then $f_1^*(s) = \lambda_1/(\lambda_1 + s)$, $f_2^*(s) = \lambda_2/(\lambda_2 + s)$. Hence the Laplace transform of the p.d.f. of $T_1 + T_2$ is

$$f_1^*(s) f_2^*(s) = \frac{\lambda_1 \lambda_2}{(\lambda_1 + s)(\lambda_2 + s)}$$

$$= \frac{\lambda_1 \lambda_2}{(\lambda_2 - \lambda_1)} \left[\frac{1}{\lambda_1 + s} - \frac{1}{\lambda_2 + s} \right].$$

But this is the Laplace transform of

$$\frac{\lambda_1 \lambda_2}{(\lambda_2 - \lambda_1)} \left[e^{-\lambda_1 t} - e^{-\lambda_2 t} \right], \lambda_2 \neq \lambda_1,$$

which is the p.d.f. of the distribution of $T_1 + T_2$, provided we accept uniqueness of Laplace transforms

The result in equation (A2.10) has an interesting alternative derivation using convolution formulae. If $B(.), C(.)$ are continuous and assume only positive arguments,

$$A(t) = \int_0^t B(t-x) C(x) dx$$

then we say $A(.)$ is the convolution of $B(.)$ with $C(.)$. In that case

$$A^*(s) = \int_0^\infty e^{-st} A(t) dt$$

$$= \int_0^\infty e^{-st} \int_0^t B(t-x) C(x) dx \, dt.$$

Interchanging the order of integration,

$$A^*(s) = \int_0^\infty \int_x^\infty e^{-st} B(t-x) C(x) dt \, dx,$$

$$A^*(s) = \int_0^\infty e^{-sx} C(x) \left[\int_x^\infty e^{-s(t-x)} B(t-x) dt \right] dx$$

$$= C^*(s)B^*(s). \tag{A2.11}$$

Example 8

If $f_n(t) = \int_0^t f_{n-1}(t-x)f(x)dx$, $n \geqslant 1$ then from (A2.11)

$$f_n^*(s) = f_{n-1}^*(s)f^*(s)$$

and, by repeated application

$$= [f^*(s)]^n, \text{ provided } f_1(.) \equiv f(.).$$

This is just a special case of a result readily obtainable from equation (A2.10).

Problem 7. If $h(t) = f(t) + \int_0^t h(t-x)f(x)dx$, where $f(t) = 0$ for $t < 0$, show

that $h^*(s) = f^*(s)/[1-f^*(s)]$. Verify that $h(t) = \sum_1^\infty f_n(t)$ where $f_n(t)$ is defined

as in Example 8. ∎

Problem 8. If $K(t) = k(t) + \int_0^t K(t-x)f(x)dx$, where all the functions are zero

for negative values of their arguments, show that

$$K(t) = k(t) + \sum_1^\infty \int_0^t k(t-x)f_n(x)dx$$

where $f_n(t)$ is defined as in Example 8. ∎

MOMENTS AND LAPLACE TRANSFORMS

The Laplace transform can only be used to find those moments which actually exist! Since

$$f^*(s) = \int_0^\infty e^{-st} f(t) dt$$

$$\frac{d}{ds}[f^*(s)] = -\int_0^\infty e^{-st} tf(t)dt.$$

Then if $\lim\limits_{s\to 0} \int_0^\infty e^{-st} tf(t)dt$ exists, it is $\int_0^\infty tf(t)dt$, the expectation of T, and

may also be evaluated as $- \lim\limits_{s\to 0} \dfrac{d}{ds} [f^*(s)]$. Similarly, if $(-1)^k \lim\limits_{s\to 0} \left[\dfrac{d^k}{ds^k} f^*(s) \right]$

exists, then it is the kth moment about the origin.

Example 9
If T has the exponential distribution with parameter λ,

$$f^*(s) = \frac{\lambda}{\lambda+s}, \frac{df^*(s)}{ds} = - \frac{\lambda}{(\lambda+s)^2}, - \lim_{s\to 0}\left[\frac{df^*(s)}{ds}\right] = \frac{1}{\lambda}.$$

Problem 9. Use the Laplace transform to find the mean and variance of a random variable which has a $\Gamma(\alpha,\lambda)$ distribution. ∎

Example 10
In this example the limiting behaviour of a function is related to that of its Laplace transform. Let $G(t) = 0$ if $t < 0$ and $g(t) = G'(t)$ be continuous.
We have from Example 5,

$$g^*(s) = -G(0) + sG^*(s),$$

and thus as $s \to 0$,

$$\lim_{s\to 0} [sG^*(s)] = g^*(0) + G(0).$$

Now providing $g(.)$ is sufficiently well behaved,

$$g^*(0) = \lim_{s\to 0} \int_0^\infty \exp(-st)g(t)dt,$$

$$= \int_0^\infty g(t)dt$$

$$= \lim_{t\to\infty} G(t) - G(0).$$

Hence, $\lim\limits_{t\to\infty} [G(t)] = \lim\limits_{s\to 0} [sG^*(s)]$.

BRIEF SOLUTIONS AND COMMENTS ON THE PROBLEMS

Problem 1

(i) $\displaystyle\int_0^\infty a\sin(bt)\exp(-st)dt = \dfrac{-a\sin(bt)\exp(-st)}{s}\Bigg]_0^\infty$

$$+\int_0^\infty \dfrac{ab\,\cos(bt)\exp(-st)}{s}\,dt,$$

$$= \dfrac{b}{s}\cdot\dfrac{as}{s^2+b^2} = \dfrac{ab}{s^2+b^2}\text{ , using Example 3.}$$

(ii) $\displaystyle\int_0^\infty \dfrac{t^n\exp(-bt)\exp(-st)}{n!}\,dt = \dfrac{1}{n!}\int_0^\infty t^n\exp[-t(b+s)]\,dt.$

After successive integration by parts, we arrive at $1/(s+b)^{n+1}$.

(iii)   Substitute $t(a+s)=ys$ in the integral.

$$\int_0^\infty \exp(-at)\cos(bt)\exp(-st)dt = \dfrac{s}{a+s}\int_0^\infty \cos\left(\dfrac{bsy}{a+s}\right)\exp(-sy)dy$$

$$= \dfrac{s}{a+s}\dfrac{s}{s^2+\dfrac{b^2\,s^2}{(a+s)^2}} = \dfrac{s+a}{[(s+a)^2+b^2]}\text{ , using Example 3.}$$

Problem 2

$$h^*(s) = \int_a^\infty g(t-a)\exp(-st)dt = \int_0^\infty g(y)\exp(-sy-as)dy$$

$$= \exp(-as)g^*(s).$$

Problem 3

$$h^*(s) = \int_0^\infty g(at)\exp(-st)dt = \int_0^\infty \dfrac{g(y)\exp(-sy/a)}{a}\,dy$$

$$= \dfrac{1}{a}g^*(s/a).$$

Problem 4

$$h^*(s) = \int_0^\infty\left[\int_0^t g(x)dx\right]\exp(-st)dt = \int_0^\infty g(x)\left[\int_x^\infty \exp(-st)dt\right]dx$$

$$= \int_0^\infty g(x)\dfrac{\exp(-sx)}{s}\,dx = \dfrac{g^*(s)}{s}\ .$$

Problem 5. After differentiating through the integral:

$$\frac{dh_n*(s)}{ds} = \int_0^\infty t^n g(t)\,(-t)\exp(-st)dt = -\int_0^\infty t^{n+1}\,g(t)\exp(-st)dt =$$

$$-h_{n+1}*(s).$$

$$h_{-1}*(s) = \int_0^\infty \frac{g(t)}{t}\exp(-st)dt, \text{ hence } \frac{d}{ds}\,[h_{-1}*(s)] = -\int_0^\infty g(t)\exp(-st)dt$$

$$= -g*(s). \text{ Integrate with respect to } s.$$

Problem 6

$$F*(s) = \int_0^\infty F(t)\exp(-st)dt = \frac{-F(s)\exp(-st)}{s}\Bigg]_0^\infty + \int_0^\infty \frac{f(t)\exp(-st)}{s}\,dt$$

$$= \frac{F(0)}{s} + \frac{f*(s)}{s}\,.$$

Problem 7

$$\int_0^\infty \left[\int_0^t h(t-x)f(x)dx\right]\exp(-st)dt$$

$$= \int_0^\infty f(x)\left[\int_x^\infty h(t-x)\exp(-st)dt\right]dx$$

$$= \int_0^\infty f(x)\exp(-sx)\left[\int_x^\infty h(t-x)\exp(-st+sx)dt\right]dx$$

$$= \int_0^\infty f(x)\exp(-sx)\,[h*(s)]\,dx = f*(s)h*(s).$$

Hence $h*(s) = f*(s) + f*(s)h*(s)$, or $h*(s) = f*(s)/[1-f*(s)] = \sum_1^\infty [f*(s)]^n$

$= \sum_1^\infty f_n*(s)$. Thus $h(t) = \sum_1^\infty f_n(t)$.

Problem 8. Take the Laplace transforms of both versions of $K(t)$. From the result in Problem 7,

$$K*(s) = k*(s) + f*(s)K*(s), \text{ or}$$

$$K*(s) = \frac{k*(s)}{1-f*(s)} = k*(s)\sum_0^\infty [f*(s)]^n = k*(s) + k*(s)\sum_1^\infty f_n*(s).$$

That is, $K(t) = k(t) + \sum_1^\infty \int_0^t k(t-x)f_n(x)dx$, from (A2.11).

Problem 9. The first derivative of $[\lambda/(\lambda+s)]^\alpha$ is $-\alpha\lambda^\alpha/(\lambda+s)^{\alpha+1} \to -\alpha/\lambda$. Hence mean is α/λ.

The second derivative is $\alpha(\alpha+1)\lambda^\alpha/(\lambda+s)^{\alpha+2} \to \alpha(\alpha+1)/\lambda^2$ as $s \to 0$. Hence variance is $\alpha(\alpha+1)/\lambda^2 - \alpha^2/\lambda^2 = \alpha/\lambda^2$.

Computing the Mean of a Distribution

Let X be a continuous, non-negative, random variable with p.d.f. $f(.)$ and c.d.f. $F(.)$. We say that the expectation of X exists if and only if the integral

$$\int_0^\infty xf(x)\,dx$$

exists. But to say that such an infinite integral exists requires

$$\lim_{t \to \infty}\left[\int_t^\infty xf(x)\,dx\right] = 0.$$

Now

$$t\,Pr[X > t] = t\int_t^\infty f(x)\,dx = \int_t^\infty tf(x)\,dx \leqslant \int_t^\infty xf(x)\,dx.$$

Hence if $E(X)$ exists, $\lim_{t \to \infty}[t\,Pr(X > t)] = 0$. But, integrating by parts,

$$\int_0^t xf(x)\,dx = -\left\{x[1-F(x)]\right\}_0^t + \int_0^t [1-F(x)]\,dx$$

$$= -t\,Pr[X > t] + \int_0^t [1-F(x)]\,dx.$$

Hence if $\lim_{t \to \infty}\left[\int_0^t xf(x)\,dx\right]$ exists then also $\lim_{t \to \infty}\int_0^t [1-F(x)]\,dx$ exists and is equal to $E(X)$.

In a similar way it can be shown that if the discrete random variable X, takes the values, $0,1,2,\ldots$ with positive probability and $E(X)$ exists, then

$$E(X) = \sum_{x=1}^\infty Pr(X \geqslant x).$$

Harder Problems

Problem 1. Record players are delivered at times which form a Poisson process of rate λ and the ith delivery consists of D_i ($\geqslant 1$) record players where $D_i (\geqslant 1)$ is a sequence of independent identically distributed integer-valued random variables with probability generating function $\pi(z)$. If $M(t)$ is the total number of record players delivered in a time interval of length t, show that the probability generating function of $M(t)$ is $\exp\{\lambda t[\pi(z)-1]\}$. Suppose that customers arrive at the retailers at time points $a, 2a, 3a, \ldots$ and each customer requires just one record player. If the retailer has a record player in stock the customer is supplied immediately. If the retailer's stock is exhausted, record players are forwarded to customers as soon as they are on hand. Supposing that initially the retailer has no record players in stock, show that the probability that the first customer is supplied immediately is $1 - e^{-\lambda a}$ and calculate the probability that the second customer is served immediately.

University of Nottingham 1974

Problem 2. A display panel of three lamps is examined at the beginning of each week by a service man who, each time, carries one spare lamp (to replace any one lamp that has failed). The lamps have independent, exponentially distributed lifetimes with mean $\dfrac{1}{\log_e 2}$ weeks. Let X_n be the number of lamps working just after the nth weekly check. Find the limiting distribution and show that its mean is equal to $\dfrac{33}{17}$.

Universities of Manchester and Sheffield 1977
(part question)

Problem 3. A stochastic process $\{X_t\}$ records the number X_t of events occurring in $(0,t)$, where

$$P\,(\text{event occurs in } (t,t + \delta t)\,|X_t \quad \text{is odd}) = \lambda\delta t + o(\delta t),$$

$$P\,(\text{event occurs in } (t,t + \delta t)\,|X_t \quad \text{is even}) = \mu\delta t + o(\delta t).$$

and, irrespective of the value of X_t, the probability of more than one event occurring in $(t,t + \delta t)$ is $o(\delta t)$, λ and μ being positive constants. Assuming that $X_0 = 0$, find the probability that X_t is even. Show that

$$E(X_t) = \frac{2\lambda\mu t}{\lambda + \mu} + \frac{(\lambda - \mu)\mu}{(\lambda + \mu)^2}\left\{e^{-(\lambda + \mu)t} - 1\right\}.$$

Imperial College London 1980

Problem 4. Consider a Poisson process of rate $\lambda(\lambda > 0)$.
(i) Let T_1 be the time to the occurrence of the first event and let $n(T_1)$ denote the number of events in the next T_1 units of time. Show that

$$E\{n(T_1)T_1\} = \frac{2}{\lambda}$$

and find $\text{var}\{n(T_1)T_1\}$.
(ii) Given that n events occur in time t, find the conditional probability density function of T_r, the time to the occurrence of the rth event $(r < n)$. Show that the conditional expectation of T_r is $rt/(n+1)$.
(iii) Consider an interval separating the occurrence of an event from the occurrence of the rth subsequent event. Find the probability distribution of the number of events of a second independent Poisson process of rate $\mu(\mu > 0)$ which occur in the interval.

Imperial College London 1980

Problem 5. A colony of living cells exists under conditions such that in any small interval of time dt, each cell has probability λdt of splitting, independent of all other cells. If a cell does split, there is probability $2/3$ that it forms two new cells, and $1/3$ that it forms three new cells. If the colony consists of just one cell at time $t = 0$ and $p_n(t)$ denotes the probability that it consists of n cells at a later time t, show that the probability generating function

$$\pi(s,t) = \sum_{n=1}^{\infty} p_n(t)s^n$$

satisfies the equation

$$\frac{\partial\pi}{\partial t} = \frac{1}{3}\lambda s(s + 3)(s - 1)\frac{\partial\pi}{\partial s}.$$

University College London 1971

Problem 6

(a)   Write down the forward differential equations for the birth and death process, and show that the steady state probability distribution is given by

$$p_n = \frac{\lambda_0 \ldots \lambda_{n-1}}{\mu_1 \ldots \mu_n} p_0 \quad n = 1, 2, \ldots .$$

(b)   A filling station has k petrol pumps and room on its forecourt for $m \geqslant k$ cars, including those at the pumps. Cars pass the station at random at a rate λ per hour, and a proportion α of these will stop and buy petrol if there is room on the forecourt. Each pump is able to dispense any grade of fuel, and service times are exponentially distributed with mean $1/\mu$ hours.

Show that in the steady state the proportion of time that the forecourt is full is

$$\frac{\rho^m}{k^{m-k} k!} \left\{ \sum_{j=0}^{k-1} \frac{\rho^j}{j!} + \frac{\rho^k}{k!} \left[\frac{1 - (\rho/k)^{m-k+1}}{1 - \rho/k} \right] \right\}^{-1}$$

where $\rho = \alpha\lambda/\mu$.

Consider the case of a filling station situated on a road for which $\lambda = 600$ cars per hour and $\alpha = 0.03$. If the average service time is 2 minutes and there is room on the forecourt for 4 cars, how many lost customers (per hour) could be saved by increasing the number of pumps from 1 to 2?

University of Southampton 1978

Problem 7. The time for which a certain type of machine will run before requiring attention by an operative has an exponential distribution with mean $1/\lambda$. The time taken by an operative to restart the machine has an exponential distribution with mean $1/\mu$. Two such machines are in use and two operatives are available to maintain them, so that a machine receives attention as soon as it breaks down. The running times and repair times for the machines are independent. Find the limiting distribution of the number of machines out of action and show that it has mean $2\rho/(1+\rho)$, where $\rho = \lambda/\mu$.

Suppose now that only one operative is available, so that a machine may have to wait for repair until the operative has completed repairing the other machine. Show that in the limiting distribution the mean number of machines out of action is $\{2\rho(1+2\rho)\}/(1+2\rho+2\rho^2)$.

Universities of Manchester and Sheffield 1977
(part question)

Problem 8. A doctor's waiting room can accommodate up to 10 patients; new arrivals leave when they find no vacancies. During each consultation period of fixed duration, $i = 0,1,2$ new patients arrive with probabilities

$$p_0 = p_1 = 0.4, \quad p_2 = 0.2.$$

At the end of each such period, the patient next in line in the waiting room is called in for consultation. If a patient arrives in the course of a period, even when the doctor is free, he is not called in until its end.

Find the generating function for the stationary probabilities of the number of patients awaiting service just after a consultation period has begun. Show that the probability of rejecting patients because the waiting room is full is smaller than $(0.1)^4$.

University of Wales, Aberystwyth 1970

Problem 9. The input to a single-server queue is according to a Poisson process of rate λ, each occurrence in the process representing the arrival of two customers. But they are served separately, the service times being independent with common probability density function $\mu e^{-\mu t}, t \geqslant 0$.

Obtain the differential equations satisfied by the probabilities for the number in the system at time t. Show that a limiting distribution can exist only if $\rho = 2\lambda/\mu < 1$, and that its probability generating function is then

$$\Pi(s) = \frac{2(1-\rho)}{2 - \rho s - \rho s^2}$$

Hence obtain the limiting mean and variance of the number in the system.

University of Leeds 1976

Problem 10. A small ferry, intended primarily for taking pedestrians across a river, can accommodate one vehicle. The vehicles that do need to cross the river in one particular direction arrive at random at rate λ, whilst the ferry leaves, in that direction, at fixed intervals of time, T minutes apart.

If the probability of n cars waiting on the river bank at the moment the ferry leaves is π_n and if $\Pi(s)$ is the corresponding probability generating function, show that

$$\Pi(s) = \frac{\pi_0 (1-s) e^{-\rho s}}{1 - s e^{\rho(1-s)}},$$

where $\rho = \lambda T$.

University of Exeter 1976

Problem 11. λ–particles and μ–particles fall on a counter in independent Poisson processes of rates 1 and 2 (per minute) respectively. On being hit by a λ–particle the counter becomes sensitive. On being hit by a μ–particle during a sensitive period the counter emits a click and becomes inert. (λ–particles hitting the counter during a sensitive period and μ–particles hitting it during an inert period have no effect.) If timing is started from a click (not to be counted) show that the expected number of clicks during the nth minute is given by

$$\frac{2}{3} + \frac{2}{9} e^{-3n} (1 - e^3).$$

(Standard results on Laplace transforms may be quoted without proof, but results specifically from renewal theory should be proved if required.)

University of Wales, Aberystwyth 1972

Problem 12. A particular system uses a number of similar components and will continue to operate as long as at least two components are still functioning. The components fail independently and each has a lifetime with p.d.f.

$$g(t) = \begin{cases} \lambda \exp\{-\lambda t\} & t \geqslant 0 \\ 0 & t < 0 \end{cases}$$

Initially, four components are installed. Subsequently, two new components are fitted on each failure of the system. Show that

$$f_1(t) = \begin{cases} 12(1 - \exp\{-\lambda t\})^2 \, \lambda \exp\{-2\lambda t\} & t \geqslant 0 \\ 0 & t < 0 \end{cases}$$

and find $f(t)$.
Show that

$$h_m(t) = \frac{6\lambda}{5} - 6\lambda \exp\{-4\lambda t\} + \frac{24}{5} \lambda \exp\{-5\lambda t\}.$$

University of Hull 1978

(Part-question: $f_1(.)$ is the p.d.f. of the time to the first fitting and $f(.)$ for the time between subsequent fittings. $h_m(t)$ is the renewal density.)

Problem 13
(a)   Let $f(x)$ be the continuous p.d.f. of operating time to failure for a particular type of component, where $f(x) = 0$ for $x < 0$, and let $X_1, X_2, \ldots, X_n$ represent independent lifetime observations for n components of this type. If $T = \sum_{i=1}^{n} X_i$, and $f^{(n)}(t)$ and $F^{(n)}(t)$ denote

the p.d.f. and c.d.f. of T respectively, show that the Laplace transforms of $F^{(n)}(t)$ and $f^{(n)}(t)$ satisfy

$$sF^{*(n)}(s) = f^{*(n)}(s)$$

and hence that

$$F^{*(n)}(s) = s^{-1}\{f^*(s)\}^n.$$

Suppose that on failure, a component is replaced immediately by a new component and let N_t denote the number of failures during $(0,t)$. Using the result that the renewal function $H(t) = E(N_t)$ may be expressed as

$$H(t) = \sum_{n=0}^{\infty} n\{F^{(n)}(t) - F^{(n+1)}(t)\},$$

show that the Laplace transform of $H(t)$ is given by

$$H^*(s) = s^{-1}f^*(s)/\{1 - f^*(s)\}.$$

(b)   A planned replacement policy is to be used in which replacements are made automatically at times $T, 2T, 3T, \ldots$, with service replacement for a failure. The cost of a service replacement is five times that of a planned replacement. If $f(x)$ is known to have the Laplace transform $f^*(s) = 2/\{(s+1)(s+2)\}$, show that

$$H(t) = \frac{2}{9}(e^{-3t} + 3t - 1),$$

and hence show that the optimum value of T satisfies the equation

$$10e^{-3T}\{1 + 3T\} = 1.$$

Brunel University 1977

Problem 14. Consider the axle of a lorry comprising a central differential with two tyres on each side. This system functions if the differential functions and at least one tyre on each side functions.

Write down the logical network for this system and obtain the structure function. Define a coherent structure: is the above system coherent?

If the tyres have independent life-times T_1, T_2, T_3, T_4 with common distribution function $P(T_i \leq t) = 1 - e^{-\lambda t}$ and the differential has life-time U (independent of the T_i) with distribution function $P(U \leq t) = 1 - e^{-\mu t}$, derive the distribution of the life-time, W, of the system and show that

$$E(W) = \frac{4}{\mu+2\lambda} - \frac{4}{\mu+3\lambda} + \frac{1}{\mu+4\lambda}.$$

University of Birmingham 1980

Problem 15

(a)   The (s,S) stock replenishment procedure is as follows. At regular intervals T replenish stock level X to S if $0 \leqslant X \leqslant s$ $(s < S)$; otherwise do nothing. Replenishments are assumed to arrive immediately. Assuming a steady state and that demand is a continuous exponential random variable on $(0,\infty)$ with parameter μ, show that the equilibrium probability distribution of stock level $X = x$ just before a review is defined by

$$f_0 = e^{-\mu S} G, \qquad x = 0;$$

and the densities

$$f_1(x) = \mu G e^{-\mu(s-x)}, \quad (0 < x \leqslant s),$$

$$f_2(x) = \mu G, \qquad\qquad (s < x < S),$$

where

$$G = \{1 + \mu(S-s)\}^{-1}$$

is the probability that $X = S$ just after review.

(b)   The costs associated with a re-order interval are as follows:

(i)   *ordering cost.* If a quantity $x > 0$ is ordered, the cost is the sum of a fixed amount C and cx, where c is constant. The costs are zero if nothing is ordered.

(ii)   *penalty per item in deficit.* If at the next re-order time demand has exceeded initial stock by an amount x, the penalty is a constant a per item in deficit.

(iii)   *storage cost per item remaining at the next re-order time.* If x items remain in stock at the next re-order time, the storage cost is a constant b times x. Show that the costs associated with a re-order interval for a level x in store just before a new order is made are:

(a) $0 \leqslant x \leqslant s$:

$$K_1(x) = C + c(S-x) + \frac{ae^{-\mu S}}{\mu} + b\left(S - \frac{1}{\mu} + \frac{e^{-\mu S}}{\mu}\right).$$

(b) $s < x < S$:

$$K_2(x) = \frac{ae^{-\mu x}}{\mu} + b\left(x - \frac{1}{\mu} + \frac{e^{-\mu x}}{\mu}\right).$$

Explain how you would find an optimal choice of s and S.

Chelsea College 1980

Problem 16. A chemical plant supplies an unstable chemical which sells at £r per tonne and costs £b per tonne to manufacture. The plant manufactures the chemical in batches of S tonnes every six weeks, this being the length of time the chemical can be stored safely. After this time, safe disposal costs a basic £d_0 plus £d_1 per tonne. Failure to meet demand, incurs contractual penalties and loss of goodwill amounting to £ℓ per tonne. If the total demand in six weeks has density $f(x)$ with mean μ show that the expected profit (£) over this period is

$$(r+d_1)\mu - (b+d_1)S - (r+d_1+\ell) \int_S^\infty (x-S)f(x)\mathrm{d}x - d_0 \int_0^S f(x)\mathrm{d}x.$$

Suppose now that the weekly demands are independent and exponentially distributed with mean 3 tonnes and that $r = 60$, $b = 10$, $d_1 = 10$, $d_0 = 0$, $\ell = 10$. Find the optimal batch size and the corresponding expected profit. You may assume, without proof, that if Y has the Gamma (λ, k) distribution with density

$$\frac{\lambda^k y^{k-1} e^{-\lambda y}}{\Gamma(k)} \;; \quad y > 0,$$

then $2\lambda Y$ is distributed as χ^2 on $2k$ degrees of freedom. If necessary use linear interpolation in the χ^2 tables provided.

University of Wales, Aberystwyth 1978

Problem 17. A farmer stores a proportion $\alpha = \dfrac{1}{3}$ of his grain every year; the annual yields $\{X_n\}$ are identically and independently distributed with mean μ and variance σ^2. Show that after a suitably large number of years, the mean and variance of his store $Z = \lim_{n\to\infty} Z_n$ will be

$$E(Z) = \frac{1}{2}\mu, \quad V(Z) = \frac{1}{8}\sigma^2.$$

Assuming the cost of storage in any year to be cZ_n, and the sale loss function to be $d\{\frac{2}{3}(Z_n + X_n) - \mu\}^2$, find the total expected annual cost in the equilibrium state. Is the proportion $\alpha = \dfrac{1}{3}$ optimal?

University of Wales, Aberystwyth 1969

Problem 18. A total of n balls is placed in two containers, A and B. Initially, a of them are in container A. Every second a ball is chosen at random and transferred from its present container to the other. If $p_r(s)$ is the probability of r balls in container A after s seconds, show that

$$n[p_r(s+1)-p_{r-1}(s)] = (r+1)p_{r+1}(s) - (r-1)p_{r-1}(s).$$

If $\phi_s(\theta)$ is the probability generating function

$$\phi_s(\theta) = \sum_{r=0}^{n} p_r(s)\theta^r,$$

show that

$$\phi_{s+1}(\theta) = \theta\,\phi_s(\theta) + \frac{1}{n}(1-\theta^2)\frac{d\phi_s(\theta)}{d\theta}.$$

By further differentiation, or otherwise, show that the expected number of balls in container A after s seconds is

$$\frac{1}{2}n + \left(a-\frac{1}{2}n\right)\left(1-\frac{2}{n}\right)^s.$$

What is the distribution of the number of balls in container A after a very large number of transfers?

University of Reading 1972

Problem 19. The non-negative integer-valued random variables $X_1, X_2, \ldots$ are independently and identically distributed with probability-generating function (p.g.f.) $P(s)$. The random variable N is distributed independently of the X_i with p.g.f. $Q(s)$. Defining

$$Y = X_1 + X_2 + \ldots + X_N,$$

prove that the p.g.f. of Y is $Q[P(s)]$.

The number of bolts, N, made by a machine during a day has p.g.f. $A(s)$. Each manufactured bolt has probability p of being defective, independently of the other bolts. Show that the number of defective bolts made during a day has the p.g.f.

$$A\{(1-p)+ps\}.$$

Denoting the mean and variance of N by μ_N and σ_N^2 respectively, show that the mean μ and variance σ^2 of the number of defective bolts made during a five-day working week are given by

$$\mu = 5p\mu_N$$
$$\sigma^2 = 5p^2\sigma_N^2 + 5p(1-p)\mu_N.$$

(Any theorems you use need not be proved but must be clearly stated.)

University of Leeds 1977

Problem 20. In a branching process with immigration, individuals of one generation independently of one another produce k offspring of like kind in the next generation, according to the probability function p_k, and, in addition, each generation is supplemented by j immigrants according to the probability function h_j, who then reproduce in the same way as the others in their generation.

Let $Q_n(s)$ be the generating function for X_n the total number in the nth generation and let $P(s) = \sum\limits_{k=0}^{\infty} p_k s^k$, $H(s) = \sum\limits_{j=0}^{\infty} h_j s^j$. Show that $Q_n(s) = H(s)Q_{n-1}(P(s))$.

Writing $P'(1) = \mu$ and $H'(1) = \nu$ show that

$$E[X_n] = \nu \frac{1 - \mu^n}{1 - \mu} + \mu^n E[X_0].$$

Suppose that the common offspring distribution is $p_0 = \alpha, p_1 = (1 - \alpha)$ and that the immigrants distribution is $h_0 = \beta, h_1 = (1 - \beta)$. If the process starts with no individuals in generation zero write down $E[X_n]$. Show that

$$Q_n \{1-(1-\alpha)^j(1-s)\} = [1-(1-\beta)(1-\alpha)^j(1-s)]\, Q_{n-1}\,\{1-(1-\alpha)^{j+1}(1-s)\}$$

and hence that the probability of having no individuals in the nth generation is

$$\prod_{k=0}^{n-1} \{1-(1-\beta)(1-\alpha)^k\}.$$

University of Leeds 1977

BRIEF SOLUTIONS AND COMMENTS ON THE PROBLEMS

Problem 1. If there are n deliveries in time t, the (conditional) p.g.f. of the number of record players delivered is $[\pi(z)]^n$. But the number of deliveries in time t has a Poisson distribution with parameter λt. Hence the (unconditional) p.g.f. of the number of record players delivered in time t is

$$G(z,t) = \sum_{n=0}^{\infty} [\pi(z)\lambda t]^n \exp(-\lambda t)/n! = e^{-\lambda t} \exp[\pi(z)\lambda t].$$

The first customer is served immediately if there has been at least one player delivered by time a. This event has probability $1-G(0,a) = 1-e^{-\lambda a}$, since $\pi(0) = 0$. The second customer is served immediately if at least two players have been delivered by time $2a$. This event has probability

$$1-G(0,2a) - \frac{\partial G}{\partial z}[(0,2a)] = 1-e^{-2a\lambda} - 2a\lambda \pi'(0) e^{-2a\lambda}.$$

Problem 2. The mean of an exponential distribution with parameter λ is $1/\lambda$. Hence $\lambda = \log 2$. $Pr[\text{life} > t_0] = \exp[-t_0 \log 2]$ and for $t_0 = 1$ (week) is $1/2$. Hence the probability of a lamp failing before the check is also $1/2$. Let $p_i^{(n)}$ be the probability that *after* the nth check exactly i lamps are functioning ($i = 1,2,3$). Since there is always one spare, $i \ne 0$.

$$p_3^{(n)} = \frac{1}{2}p_3^{(n-1)} + \frac{1}{4}p_2^{(n-1)}$$

$$p_2^{(n)} = \frac{3}{8}p_3^{(n-1)} + \frac{1}{2}p_2^{(n-1)} + \frac{1}{2}p_1^{(n-1)}$$

$$p_1^{(n)} = \frac{1}{8}p_3^{(n-1)} + \frac{1}{4}p_2^{(n-1)} + \frac{1}{2}p_1^{(n-1)}.$$

Suppose $p_i^{(n)} \to p_i$, then $p_2 = 8p_1/5, p_3 = 4p_1/5$. Since $p_1 = 1-p_2-p_3$, we get $p_1 = 5/17, p_2 = 8/17, p_3 = 4/17$ and $E(X) = \Sigma np_n = 33/17$ as required.

Problem 3. Let $p_E(t)$ be the probability that X_t is even

$$p_E(t+\delta t) = p_E(t)(1-\mu\delta t) + [1-p_E(t)]\lambda\delta t + o(\delta t)$$

$$p_E'(t) \quad = -(\lambda + \mu)p_E(t) + \lambda. \tag{1}$$

Multiply by the integrating factor $\exp[(\lambda + \mu)t]$

$$\frac{d}{dt}\left\{\exp[(\lambda + \mu)t]\,p_E(t)\right\} = \lambda \exp[(\lambda + \mu)t],$$

using the initial condition $p_E(0) = 1$,

$$p_E(t) = \frac{\lambda}{\lambda + \mu} + \frac{\mu}{\lambda + \mu} \exp[-(\lambda + \mu)t]. \tag{2}$$

The rider is (perhaps) most safely tackled by computing $\frac{d}{dt}[E(X_t)]$. Similar arguments as for (1) provide

$$p_{2n}'(t) \quad = -\mu p_{2n}(t) + \lambda p_{2n-1}(t), \tag{3}$$

$$p_{2n-1}'(t) \quad = -\lambda p_{2n-1}(t) + \mu p_{2n-2}(t). \tag{4}$$

From (3), (4) obtain

$$\frac{d}{dt}[E(X_t)] = \lambda\, Pr[X_t \text{ is odd}] + \mu\, Pr[X_t \text{ is even}].$$

Substitute for $p_E(t)$ from (2) and obtain the result by integration.

Problem 4

(i)   Given $T_1 = t_1$, the number of events in the next t_1 units of time has a Poisson distribution with parameter λt_1. Hence $E[n(T_1)T_1|T_1 = t_1] = E[n(t_1)t_1|T_1 = t_1] = t_1 E[n(t_1)|T_1 = t_1] = t_1 \lambda t_1 = \lambda t_1{}^2$. The time to the first event has an exponential distribution with parameter λ. Hence $E(\lambda T_1{}^2) = \lambda(2!/\lambda^2) = 2/\lambda$.

$$V[n(T_1)T_1] = E[\{n(T_1)T_1\}^2] - [E\{n(T_1)T_1\}]^2$$

$$= E[T_1^2\{n(T_1)\}^2] - \frac{4}{\lambda^2}.$$

Now given $T_1 = t_1$, $E[\{n(t_1)\}^2|T_1 = t_1] = \lambda t_1 + \lambda^2 t_1{}^2$.

Hence $V[n(T_1)T_1] = E(\lambda T_1{}^3 + \lambda^2 T_1{}^4) - \dfrac{4}{\lambda^2}$

$$= \lambda\frac{3!}{\lambda^3} + \lambda^2\frac{4!}{\lambda^4} - \frac{4}{\lambda^2} = \frac{26}{\lambda^2}.$$

(ii)   Some readers will recognise at once that the distribution of T_r is that of the rth order statistic in a random sample of n observations from a uniform distribution over the interval $(0,t)$, whence the answer is immediate. More laboriously consider y such that $0 < y < t$. There will k events in $(0,y)$ and $n-k$ in (y,t), given that there are n in $(0,t)$ with conditional probability

$$\frac{\{[(\lambda y)^k \exp(-\lambda y)]/k!\}\ \{[(\lambda t - \lambda y)^{n-k} \exp(-\lambda t + \lambda y)]/(n-k)!\}}{[(\lambda t)^n \exp(-\lambda t)]/n!}$$

$$= \frac{n!}{k!\,(n-k)!} \cdot \frac{y^k(t-y)^{n-k}}{t^n}.$$

The *distribution* function of T_r, $F_r(y)$, is the conditional probability that there are at least r events in $(0,y)$. Hence

$$F_r(y) = \sum_{k=r}^{n} \binom{n}{k}[y^k(t-y)^{n-k}/t^n].$$

$$\int_0^t [1 - F_r(y)]\,dy = t - (n-r+1)t/(n+1) = rt/(n+1).$$

(iii)   The time T between an event and the rth subsequent event has a $\Gamma(r,\lambda)$ distribution. Given that $T = t$, the number of events in the second process has a Poisson distribution with parameter μt. Hence the unconditional probability of n events in the second process is

$$\int_0^\infty \frac{(\mu t)^n \exp(-\mu t)}{n!} \frac{\lambda (\lambda t)^{r-1} \exp(-\lambda t)}{(r-1)!} \, dt$$

$$= \binom{n+r-1}{r-1} \left(\frac{\mu}{\lambda+\mu}\right)^n \left(\frac{\lambda}{\lambda+\mu}\right)^r, n = 0,1,2,\ldots.$$

Problem 5

$$p_n(t+\delta t) = p_n(t) (1-n\lambda\delta t) + \frac{2}{3} p_{n-1}(t) (n-1) \lambda\delta t + \frac{1}{3} p_{n-2}(t) (n-2) \lambda\delta t,$$

$$n \geqslant 3.$$

$$\frac{dp_n(t)}{dt} = -n\lambda p_n(t) + \frac{2(n-1)\lambda}{3} p_{n-1}(t) + \frac{(n-2)\lambda}{3} p_{n-2}(t).$$

The last equation is also satisfied for $n = 1,2$, since $p_{-1}(t) \equiv 0$. Hence multiply by s^n and sum over n to obtain result.

Problem 6. The first part is book-work and p_n, λ_n, μ_n are to be taken to have the same meaning as in Chapter 3, section 3.6.

We have $\lambda_n = \alpha\lambda$, $0 \leqslant n \leqslant m-1$, $\lambda_n = 0$, $n \geqslant m$;

$$\mu_n = n\mu, \; 1 \leqslant n \leqslant k, \qquad \mu_n = k\mu, \quad n > k.$$

Using the formula in (a),

$$p_n = \frac{(\alpha\lambda)^n}{(k\mu)^{n-k}k!\mu^k} p_0, \quad k \leqslant n \leqslant m;$$

$$p_n = \frac{(\alpha\lambda)^n}{n!\mu^n} p_0, \qquad 1 \leqslant n \leqslant k-1.$$

To find p_0, use the condition

$$p_0 + \sum_{i=1}^{k-1} p_i + \sum_{i=k}^{m} p_i = 1,$$

and hence calculate p_m, which is the required result. Calculate p_4 for the two different values of k and bear in mind that not all the cars passing a full fore-court would otherwise have used the station.

Problem 7. Let $\phi_i(t)$, be the probability that i machines are functioning at time t $(i = 0,1,2)$. When only one operator is available:

(i) $\phi_2(t+\delta t) = \phi_2(t) (1-\lambda\delta t)^2 + \phi_1(t) (1-\lambda\delta t)\mu \, \delta t.$

(ii) $\phi_1(t+\delta t) = \phi_2(t)2\lambda\delta t + \phi_1(t) (1-\lambda\delta t) (1-\mu\delta t) + \phi_0(t)\mu\delta t.$

(iii) $\phi_0(t+\delta t) = \phi_0(t) (1-\mu\delta t) + \phi_1(t)\lambda\delta t(1-\mu\delta t).$

In the stationary distribution, from (iii), we have

$$\phi_1 = \mu\phi_0/\lambda;$$

and from (i), $\phi_2 = \tfrac{1}{2}\mu^2\phi_0/\lambda^2$. Use the constraint $\phi_0 + \phi_1 + \phi_2 = 1$ to solve for ϕ_1 and ϕ_2. The expected number of machines *working* is $\dfrac{2(\rho+1)}{2\rho^2+2\rho+1} = 2 -$ expected number out of action. When there are two repair men, it has to be remembered that when *both* are working, the probability that one finishes in $(t,t+\delta t)$ is $2\mu\delta t + o(\delta t)$.

If $\psi_n(t)$ is the probability that i machines are functioning at time $t (i = 0,1,2)$

$$\psi_2(t+\delta t) = \psi_2(t)(1-\lambda\delta t)^2 + \psi_1(t)(1-\lambda\delta t)\mu\delta t.$$

$$\psi_1(t+\delta t) = \psi_2(t)2\lambda\delta t + \psi_1(t)(1-\lambda\delta t)(1-\mu\delta t) + \psi_0(t)2\mu\delta t.$$

$$\psi_0(t+\delta t) = \psi_0(t)(1-\mu\delta t)^2 + \psi_1(t)\lambda\delta t(1-\mu\delta t).$$

Problem 8. There cannot be 10 patients in the waiting room just *after* a consultation period has begun. Let π_i $(i = 0,1,2,\ldots,9)$ be the required stationary probability of i patients in the waiting room. By considering two successive periods, we have, after a large number of periods,

$$\pi_9 = \pi_8 p_2 + \pi_9 p_1 + \pi_9 p_2. \tag{1}$$

$$\pi_n = \pi_{n+1}p_0 + \pi_n p_1 + \pi_{n-1}p_2, \quad 1 \leqslant n \leqslant 8. \tag{2}$$

From equation (1), $\pi_9(1-p_1-p_2) = \pi_9 p_0 = \pi_8 p_2$.
Set $n = 8$ in (2) and obtain

$$\pi_8 = \pi_9 p_0 + \pi_8 p_1 + \pi_7 p_2$$

$$= \pi_8 p_2 + \pi_8 p_1 + \pi_7 p_2,$$

or $\pi_8 p_0 = \pi_7 p_2$. By taking $n = 7,6,\ldots,1$ we obtain $\pi_{n+1}p_0 = \pi_n p_2, n = 0,1,2,\ldots,8$. But $\sum_{i=0}^{9} \pi_i = 1$, hence

$$\pi_0 \sum_{i=0}^{9} (p_2/p_0)^i = 1 \Rightarrow \pi_0 = 1/[2-(1/2)^9],$$

and $\pi_r = \pi_0/2^r, r = 1,2,\ldots,9$.
The probability of the waiting room being full is $\pi_9 p_2 = (0.2)/(2^{10}-1)$. The probability that an arriving patient finds the room full is $(\pi_9 p_2)/2 < (0.1)^4$.

Problem 9. The equations are obtained by the usual arguments together with the consideration that if there is an arrival, this involves *two* customers. For $n \leqslant 2$,

$p_n(t+\delta t) = p_n(t)\,[1-(\lambda+\mu)\delta t] + p_{n-2}(t)\lambda\delta t + p_{n+1}(t)\mu\delta t + o(\delta t).$

$$p_n'(t) = -(\lambda+\mu)p_n(t) + \lambda p_{n-2}(t) + \mu p_{n+1}(t). \tag{1}$$

$$p_1'(t) = -(\lambda+\mu)p_1(t) + \mu p_2(t). \tag{2}$$

$$p_0'(t) = -\lambda p_0(t) + \mu p_1(t). \tag{3}$$

In the stationary distribution, set $p_n'(t) = 0$, $p_n(t) = p_n$. Multiply equation (1) by s^n and sum from 1 to ∞ (note $p_{-1}(t) \equiv 0$).

$$0 = -(\lambda+\mu)\,[\pi(s) - p_0] + \lambda s^2 \pi(s) + \mu[\pi(s) - p_0 - sp_1]/s. \tag{4}$$

From (3), we have $\lambda p_0 = \mu p_1$ and substituting for μp_1 in (4) and cancelling $(s-1)$, we obtain,

$$\pi(s)\,[\mu-\lambda s(s+1)] = \mu p_0.$$

After setting $s = 1$, we obtain $p_0 = (1 - 2\lambda/\mu)$ and hence result. The mean is found from $\pi'(1)$, the variance from $\pi''(1) + \pi'(1) - [\pi'(1)]^2$.

Problem 10. Let $p_n^{(r)}$ be the probability that there are n cars left behind as the ferry leaves for the rth trip and ϕ_i be the probability that i cars arrive in time T. There will be $n(\geqslant 1)$ if:
(a) there were $n-k$ left after trip $(r-1)$ commenced, $k+1$ arrived in time T and one was loaded for trip r. ($0 \leqslant k \leqslant n$).
(b) there were $n+1$ left after trip $(r-1)$ commenced, no additional car arrived in time T and one was loaded for trip r. Hence

$$p_n^{(r)} = \sum_{k=0}^{n} p_{n-k}^{(r-1)}\phi_{k+1} + p_{n+1}^{(r-1)}\phi_0. \tag{1}$$

In the limit, as $r \to \infty$, let $p_n^{(r)} \to \pi_n$, then from equation (1)

$$\pi_n = \sum_{k=0}^{n} \pi_{n-k}\phi_{k+1} + \pi_{n+1}\phi_0. \tag{2}$$

The case $n = 0$ is slightly different. There are three possibilities:
(c) there was one left after trip $r-1$, this was loaded for trip r and no car arrived in time T;
(d) there was none left after trip $r-1$, one arrived in time T and this was loaded;
(e) there was none left after trip $r-1$, none arrived in time T; hence

$$p_0^{(r)} = p_1^{(r-1)}\phi_0 + p_0^{(r-1)}\phi_1 + p_0^{(r-1)}\phi_0, \tag{3}$$

$$\pi_0 = \pi_1\phi_0 + \pi_0\phi_1 + \pi_0\phi_0. \tag{4}$$

Now form the generating function $\Pi(s) = \sum_{n=0}^{\infty} \pi_n s^n$ by multiplying equation

(2) by s^n, summing over $n \geqslant 1$ and using equation (4). Beware of $\sum\limits_{n=1}^{\infty} \sum\limits_{k=0}^{n}$ $\pi_{n-k}\phi_{k+1}s^n$ in which the term $k=0$ must be split off before interchanging the order of summation. The number of arrivals in time T has a Poisson distribution with parameter λT, hence $\phi_i = (\lambda T)^i \exp(-\lambda T)/i!$.

Problem 11. A click is only obtained if a λ–particle is followed by a μ–particle. Thus we have an ordinary renewal process for which the time between events has a distribution which is the sum of independent exponential distributions with parameters 1,2. The p.d.f. of the sum of two such distributions has Laplace transform $2/[(1+s)(2+s)]$. Hence the Laplace transform of the renewal density function of the process is $\{2/[(1+s)(2+s)]\}/\{1-2/[(1+s)(2+s)]\} = 2/[s(s+3)]$ $= (2/3s) - [2/(3s+9)]$. Hence the renewal density function, $h(t)$, is $\dfrac{2}{3}$ $[1-\exp(-3t)]$ and the expected number of clicks in $(0,t)$ is

$$\frac{2}{3} t + \frac{2}{9} [\exp(-3t)-1] = H(t).$$

Now evaluate $H(n) - H(n-1)$.

Problem 12. The system will require a first fitting before t if at least 3 components have failed by t. This has probability

$$\binom{4}{3}(1-e^{-\lambda t})^3 e^{-\lambda t} + (1-e^{-\lambda t})^4.$$

This is the distribution function and should be differentiated with respect to t to obtain

$$f_1(t) = 12(1-e^{-\lambda t})^2 \lambda e^{-2\lambda t} = 12\lambda(e^{-2\lambda t} - 2e^{-3\lambda t} + e^{-4\lambda t}).$$

Hence $\quad f_1^*(s) = 12\lambda\left(\dfrac{1}{s+2\lambda} - \dfrac{2}{s+3\lambda} + \dfrac{1}{s+4\lambda}\right)$

$$= \frac{24\lambda^3}{(s+2\lambda)(s+3\lambda)(s+4\lambda)}.$$

After the initial fitting, there will be 3 components and the system will fail, in a subsequent interval t, if at least 2 of these have failed. This has probability

$$\binom{3}{2}(1-e^{-\lambda t})^2 e^{-\lambda t} + (1-e^{-\lambda t})^3.$$

Hence $f(t) = 6(1-e^{-\lambda t})\lambda e^{-2\lambda t} = 6\lambda(e^{-2\lambda t} - e^{-3\lambda t})$ and

$$f^*(s) = 6\lambda\left(\frac{1}{s+2\lambda} - \frac{1}{s+3\lambda}\right) = \frac{6\lambda^2}{(s+2\lambda)(s+3\lambda)}.$$

The Laplace transform of the (modified) renewal density function is $h_m^*(s) = f_1^*(s)/[1-f^*(s)] = 24\lambda^3/[s(s+4\lambda)(s+5\lambda)]$. Now find the Laplace transform of the displayed $h_m(t)$.

Problem 13. (b) $h^*(s) = \left(\dfrac{1}{s} - \dfrac{1}{s+3}\right)\dfrac{2}{3}$.

Hence

$$h(t) = \frac{2}{3} - \frac{2}{3}\exp(-3t), \text{ and thus}$$

$$H(t) = \frac{2}{3}t + \frac{2}{9}[\exp(-3t)-1].$$

Consider any period of length T which begins with a planned service. Hence the expected cost *per unit time* is

$$\frac{c + 5cH(T)}{T}$$

and should be minimised with respect to T to obtain the optimum value.

Problem 14. The probability that both tyres fail before time t on one side is p_1^2, where $p_1 = 1-\exp(-\lambda t)$. Probability that at least one tyre functions on one side is $1-p_1^2$. Hence at least one functions on both sides with probability $(1-p_1^2)^2$. Probability that differential is still functioning at time t is $\exp(-\mu t) = 1-p_2$. Hence probability that system is not functioning at time t is $1-(1-p_1^2)^2(1-p_2) = F(t) = 1-[4\exp(-\mu t-2\lambda t) - 4\exp(-\mu t-3\lambda t) + \exp(-\mu t-4\lambda t)]$. This is the distribution function of the time to failure. Hence the mean time to failure is

$$\int_0^\infty [1-F(t)]\,dt.$$ A system is coherent if its structure function is monotone and

every component is relevant.

Problem 15. Let X_n be the stock level immediately before the nth review and Z the consumption since the $(n-1)$th review. Suppose $f_{0,n} = Pr[X_n = 0]$ and $f_{1,n}(.), f_{2,n}(.)$ are the densities for $0 < X_n \leqslant s$, and $s < X_n < S$, respectively. We calculate $Pr[X_n > x]$ when $x > s$. Now $X_n > x$ can occur from the following situations at the $(n-1)$th review.
(a) $X_{n-1} > x$, there was thus no replenishment, and the subsequent demand was such that $X_{n-1}-Z > x$. But this requires $Z < X_{n-1}-x$. For any particular value x_{n-1}, the $Pr[Z < x_{n-1}-x] = 1-\exp[-\mu(x_{n-1}-x)]$. Hence the total contribution for $X_{n-1} > x$ is

$$\int_x^S [1-\exp\{-\mu(x_{n-1}-x)\}]f_{2,n-1}(x_{n-1})\,dx_{n-1}. \tag{1}$$

(b) $0 < X_{n-1} < s$, the stock was replenished to level S and the demand was such that $S-Z > x$ or $Z < S-x$. The total contribution in probability is

$$\int_0^s [1-\exp\{-\mu(S-x)\}]f_{1,n-1}(x_{n-1})dx_{n-1}. \tag{2}$$

(c) $X_{n-1} = 0$, the stock was replenished to S, $Z < S-x$, with total probability,

$$[1-\exp\{-\mu(S-x)\}]\, Pr(X_{n-1} = 0). \tag{3}$$

But the sum of the probabilities in equations (1), (2), (3) is $Pr(X_n > x)$, when $x > s$. Hence

$$Pr(X_n > x) = Pr(X_{n-1} > x) - e^{\mu x} \int_x^S f_{2,n-1}(x_{n-1})\exp(-\mu x_{n-1})dx_{n-1}$$

$$+ [1-\exp\{-\mu(s-x)\}]\, [1-Pr(X_{n-1} > s)].$$

Now in the equilibrium distribution, as $n \to \infty$

$$Pr(X_n > x) \to Pr(X_{n-1} > x) \text{ and}$$

$$f_{2,n-1}(.) \to f_2(.).$$

Taking the limit as $n \to \infty$ in equation (4),

$$0 = - e^{\mu x} \int_x^S f_2(y)e^{-\mu y}dy + [1-e^{-\mu(S-x)}]\, [1-Pr(X > s)]. \tag{5}$$

To recover $f_2(.)$ from equation (5), differentiate with respect to x, when we obtain

$$0 = -\mu e^{\mu x} \int_x^S f_2(y)e^{-\mu y}dy + f_2(x) - \mu e^{-\mu(S-x)}[1-Pr(X > s)]. \tag{6}$$

Substituting from equation (5) into (6), and collecting terms, we arrive at

$$f_2(x) = \mu\,[1-Pr(X > s)] = \mu Pr(X \leqslant s).$$

This of the form μG as required by the question. Similarly we may find $f_1(.)$ and f_0, and G from the usual normalising condition.

For the last part, if $X \leqslant s$ an order for $S - x$ is placed at cost $C + c(S-x)$. If Z is the demand, the expected holding charge is

$$b \int_0^S (S-z)\mu e^{-\mu z}dz$$

$$= b\left(S - \frac{1}{\mu} + \frac{e^{-\mu S}}{\mu}\right).$$

The expected shortage penalty is

$$a \int_S^\infty (z-S)\mu e^{-\mu z} dz$$

$$= \frac{ae^{-\mu S}}{\mu}.$$

By adding these costs we obtain $K_1(x)$. If $s < x < S$, then there is no replenishment and S is clearly replaced by x in the expected holding and shortage costs.

Allowing for random variation in the level just before a new order is made, the overall expected cost for the policy is

$$K_1(0)Pr(X=0) + \int_0^s K_1(x)f_1(x)dx + \int_s^S K_2(x)f_2(x)dx.$$

This last expression should be minimised with respect to s and S.

Problem 16. Suppose the demand is x.

(i) If $x < S$, the gain is rx, $S-x$ is left over and, for disposal, costs $d_0 + (S-x)d_1$.

(ii) If $x \geqslant S$, the gain is rS, $x-S$ is unfulfilled demand, with a goodwill loss of $(x-S)\ell$.

In either case the cost of the material is bS. Hence expected profit is

$$\int_0^S rxf(x)dx + Sr \int_S^\infty f(x)dx$$

$$- \int_0^S [d_0 + (S-x)d_1] f(x)dx - \ell \int_S^\infty (x-S)f(x)dx - bS.$$

Using obvious equalities of the type,

$$r \int_0^S xf(x)dx = r \left[\int_0^\infty xf(x)dx - \int_S^\infty xf(x)dx \right]$$

$$= r\mu - r \int_S^\infty xf(x)dx,$$

we obtain the desired result. Differentiate with respect to S, and the optimal batch size S^*, satisfies

$$-(b+d_1) - (r+d_1+\ell) [F(S^*)-1] - d_0f(S^*) = 0,$$

where $F(.)$ is the c.d.f. of the demand. With the values given, $F(S^*) = 3/4$. Noting that the total demand for six weeks must have a $\Gamma(1/3,6)$ distribution, use χ_{12}^2 distribution to find approximate value of S^*.

Problem 17. Let the store's content be Z_n at the end of year n and the yield during the year $(n,n+1)$ be X_n. Then

$$Z_{n+1} = \alpha(Z_n + X_n)$$

$$E(Z_{n+1}) = \alpha[E(Z_n) + E(X_n)].$$

In the limit, $E(Z) = \alpha E(Z) + \alpha\mu$. If $\alpha = 1/3$, $E(Z) = \mu/2$.

$$V(Z_{n+1}) = \alpha^2 V(Z_n) + \alpha^2 \sigma^2.$$

In the limit, $V(Z) = \alpha^2 V(Z) + \alpha^2 \sigma^2$. If $\alpha = 1/3$, $V(Z) = \sigma^2/8$. At the end of year $n+1$, $(1-\alpha)(Z_n+X_n) = (1-\alpha)Z_{n+1}/\alpha \to (1-\alpha)Z/\alpha$, is sold. Hence the total expected annual cost in the equilibrium state is

$$d E \left(\frac{1-\alpha}{\alpha}Z - \mu\right)^2 + cE(Z)$$

$$= d V\left(\frac{1-\alpha}{\alpha}Z\right) + \frac{c\alpha\mu}{1-\alpha} = d\frac{(1-\alpha)^2}{\alpha^2}\frac{\alpha^2\sigma^2}{1-\alpha^2} + \frac{c\alpha\mu}{1-\alpha}$$

$$= d\frac{(1-\alpha)\sigma^2}{1+\alpha} + \frac{c\alpha\mu}{1-\alpha}.$$

This expression has derivative $-\dfrac{2d\sigma^2}{(1+\alpha)^2} + \dfrac{c\mu}{(1-\alpha)^2}$ with respect to α. Hence if $c\mu > 2d\sigma^2$, the slope is positive at $\alpha = 0$ and the optimal choice is $\alpha = 0$. If $c\mu < 2d\sigma^2$, the optimal value satisfies $c\mu/(1-\alpha)^2 = 2d\sigma^2/(1+\alpha)^2$.

Problem 18. We can obtain r at time $(s+1)$ from

$$r-1 \text{ at } s \text{ with probability } \frac{n-r+1}{n}$$

or $\qquad r+1 \text{ at } s \text{ with probability } \dfrac{r+1}{n}$

$$p_r(s+1) = p_{r-1}(s)\frac{n-r+1}{n} + p_{r+1}(s)\frac{r+1}{n}, r = 0,1,\ldots,n \text{ with the proviso that}$$
$p_r(s) = 0$ when $r > n$ or $r < 0$.
Therefore

$$n[p_r(s+1) - p_{r-1}(s)] = (r+1)p_{r+1}(s) - (r-1)p_{r-1}(s). \tag{1}$$

Multiply (1) by θ^r and sum over $r = 0$ to $r = n$.

$$n\phi_{s+1}(\theta) - n\theta[\phi_s(\theta) - \theta^n p_n(s)] = \frac{d}{d\theta}[\phi_s(\theta)]$$

$$- \theta^2\frac{d}{d\theta}[\phi_s(\theta) - \theta^n p_n(s)],$$

i.e. $\phi_{s+1}(\theta) = \theta\,\phi_s(\theta) + \frac{1}{n}(1-\theta^2)\frac{d\phi_s(\theta)}{d\theta}$.   (2)

Differentiating (2),

$$\frac{d\phi_{s+1}(\theta)}{d\theta} = \phi_s(\theta) + \theta\frac{d\phi_s(\theta)}{d\theta} - \frac{2\theta}{n}\frac{d\phi_s(\theta)}{d\theta} + \frac{1}{n}(1-\theta^2)\frac{d^2\phi_s(\theta)}{d\theta^2}$$

Put $\theta = 1$ and let μ_{s+1}, μ_s be the mean values.

$$\mu_{s+1} = 1 + (1 - 2/n)\mu_s.$$

Solving this recurrence equation,

$$\mu_s = n/2 + (a - \tfrac{1}{2}n)(1 - 2/n)^s.$$

In the limit as $s \to \infty$ we get, from (2),

$$\phi(\theta) = \frac{1}{n}(1 + \theta)\frac{d\phi}{d\theta},$$

hence $\phi(\theta) = K(1 + \theta)^n$.

When $\theta = 1, \phi(\theta) = 1$ so that $K = \dfrac{1}{2^n}$

Therefore $\phi(\theta) = \left(\dfrac{1+\theta}{2}\right)^n$.

This is the p.g.f. of a binomial distribution. It is as if each ball were placed with probabilities ½, ½ in one or other of the two containers.

Problem 19. The (conditional) p.g.f. of Y given $N = n$ is

$$E(s^Y) = E\left(s^{\sum_1^n X_i}\right) = \prod_1^n E(s^{X_i}) = \prod_1^n P(s) = [P(s)]^n.$$

Hence the (unconditional) p.g.f. of Y is

$$\Sigma[P(s)]^n Pr(N = n) = Q[P(s)].$$

Let $X_i = 1$ if the ith bolt is defective and $X_i = 0$ if it is not. Then $\sum_1^N X_i$ is the number of defective bolts produced per day. Now $P(s) = (1-p) + ps$. Hence $A[P(s)] = A[(1-p)+ps)]$ and the mean number of defective bolts is $\left\{\dfrac{d}{ds}[A(1-p+ps)]\right\}_{s=1} = A'(1)p = \mu_N p$. Hence $5p\mu_N$ for five days. After a second differentiation, we obtain

$$A''(1)p^2 = (\sigma_N^2 + \mu_N^2 - \mu_N)p^2.$$

But

$$[A''\,(1-p+ps)]_{s=1}$$

is also

$$V(Y) + E^2(Y) - E(Y).$$

Hence

$$V(Y) = p^2\,\sigma_N^2 + p(1-p)\mu_N, \text{ and } \sigma^2 = 5p^2\,\sigma_N^2 + 5p(1-p)\mu_N \text{ for five days.}$$

Problem 20. Let X_n be the total number in generation n, Y_i the number of offspring of the ith member of generation $n-1$, and I_n the number of immigrants into the nth generation. Then clearly,

$$X_n = Y_1 + Y_2 + \ldots + Y_{X_{n-1}} + I_n. \text{ Hence,}$$

$$Q_n(s) = E\left(s^{X_n}\right) = E\left(s^{Y_1 + Y_2 + \ldots + Y_{X_{n-1}}}\right) E(s^{I_n}).$$

Now given $X_{n-1} = x_{n-1}$, the conditional p.g.f. of $Y_1 + Y_2 \ldots + Y_{X_{n-1}}$ is $[P(s)]^{X_{n-1}}$. Hence

$$Q_n(s) = H(s)E\,[P(s)]^{X_{n-1}} = H(s)Q_{n-1}\,[P(s)].$$

$$E(X_n) = Q_n'(1) = Q_{n-1}'\,[P(1)]\,P'(1)H(1) + Q_{n-1}\,[(1)]\,H'(1)$$

$$= E(X_{n-1})\mu + \nu, \text{ that is}$$

$$\frac{E(X_r)}{\mu^r} - \frac{E(X_{r-1})}{\mu^{r-1}} = \frac{\nu}{\mu^r}\,.$$

Summing this last equation from $r = 1$ to $r = n$, we obtain the required result.

If

$$P(s) = \alpha + s(1-\alpha), H(s) = \beta + s(1-\beta),$$

then

$$Q_n(s) = [\beta + s(1-\beta)]\,Q_{n-1}\,[\alpha + s(1-\alpha)].$$

That is,

$$Q_n\,[1-(1-\alpha)^j(1-s)] = [1-(1-\beta)(1-\alpha)^j(1-s)]\,Q_{n-1}\,[1-(1-\alpha)^{j+1}(1-s)].$$

Now the probability of no individuals in the nth generation is $Q_n(0)$. But

$$Q_n(0) = \beta\,Q_{n-1}(\alpha)$$

$$= [1-(1-\beta)]\,Q_n\,[1-(1-\alpha)]$$

$$= [1-(1-\beta)]\,[1-(1-\beta)(1-\alpha)]\,Q_{n-2}\,[1-(1-\alpha)^2\,]$$

$$= [1-(1-\beta)]\,\ldots\,[1-(1-\beta)(1-\alpha)^{n-2}]\,Q_1\,[1-(1-\alpha)^{n-1}\,].$$

But $Q_1(s) = 1-(1-\beta)(1-s)$, hence

$$Q_1\left[1-(1-\alpha)^{n-1}\right] = 1-(1-\beta)\left[1-\left\{1-(1-\alpha)^{n-1}\right\}\right]$$
$$= 1-(1-\beta)(1-\alpha)^{n-1},$$

and the result follows.

Index